A Practical Introduction to Applied Statistics for Materials Scientists and Engineers

David J. Keffer
Professor of Materials Science & Engineering
University of Tennessee
Knoxville, TN

Copyright © 2015 David J. Keffer
All rights reserved.

First Edition

ISBN-13: 978-1517521936

This book is dedicated to my colleagues in the Materials Science & Engineering Department at the University of Tennessee, Knoxville, who provided a good home in which this book could be written.

Preface

Materials Scientists and Engineers work in a quantitative field in which the mechanical, thermodynamic, rheological and electronic properties of the materials are described with numbers. Therefore, a basic proficiency in applied statistics is necessary to be able to interpret and communicate the significance and reliability of these numbers. The numbers provide the technical basis upon which are made engineering decisions that impact the successful deployment of the material and ultimately the success of the project, including many elements such as public safety. Understanding the statistical confidence in these numbers is therefore of practical significance.

This book provides a practical but fundamental mathematical understanding of applied statistics and the theory of probability that underlies it. This book is targeted at undergraduates. It is used in a course during the third year that covers both applied statistics and numerical methods. Thus, the contents herein are intended to cover half of one semester. The content is divided into six chapters (or modules), covering (1) probability, (2) random variables, (3) expectations, (4) discrete distributions, (5) continuous distributions and (6) sampling and estimation. Other important statistical procedures, such as regression, are covered in the second half of the course, numerical methods.

The philosophy espoused in this book is to equip the student with a compact but broadly applicable set of practical problem-solving tools such that the student emerges at the end of the course with the ability to clearly articulate the statistical reliability of the numbers used to describe the materials of interest.

Finally, we live in a word where the importance of literacy, the ability to make sense of words, is taken for granted. At the same time, importance of numeracy, the ability to make sense of numbers, is neglected. In fact, poor numeracy skills of the general public (and its elected leaders) are regularly exploited, both intentionally and unintentionally, particularly by those who seek to sway public opinion. An introduction to applied statistics is the first step toward the ability to extract meaning and significance from raw data. In this book, the application of these techniques is toward materials science. However, the broader applicability of such techniques is a natural consequence of one equipped with these skills.

Summary of the Contents of this Book

This book covers:

- an introduction to probability
- random variables and probability distribution functions
- expectations
- common discrete probability distribution functions
- common continuous probability distribution functions
- sampling, estimation and confidence intervals

This book does not cover:

- regression (included in the second half of the course)
- analysis of Variance (ANOVA)
- countless other more advanced statistical topics

Table of Contents

Preface iv
Summary of the Contents of this Book v
Table of Contents vi
List of Subroutines ix

Chapter 1. Probability 2
    1.1. Introduction 2
    1.2. Vocabulary 2
    1.3. Counting Rules 6
    1.4. Probability 11
    1.5 Subroutines 23
    1.6. Problems 24

Chapter 2. Random Variables and Probability Distributions 26
    2.1. Introduction 26
    2.2. Random Variables & Sample Spaces 26
    2.3. Discrete Probability Distribution Functions (PDFs) 27
    2.4. Continuous Probability Density Functions (PDFs) 31
    2.5. Relations between Inequalities 35
    2.6. Discrete Joint Probability Distribution Functions 37
    2.7. Continuous Joint Probability Density Functions 38
    2.8. Marginal Distributions and Conditional Probabilities 40
    2.9. Statistical Independence 44
    2.10. Problems 45

Chapter 3. Expectations 47
3.1. Introduction 47
3.2. Mean of a Random Variable.. 47
3.3. Mean of a Function of a Random Variable 48
3.4. Mean of a Function of Two Random Variables 50
3.5. Variance of a Random Variable 51
3.6. Standard Deviation 54
3.7. Variance of a Function of a Random Variable 55
3.8. Variance & Covariance of a Function of Two Random Variables..................... 56
3.9. Correlation Coefficients 58
3.10. Means and Variances of linear combinations of Random Variables................. 60
3.11. An Extended Example for Discrete Random Variables 64
3.12. An Extended Example for Continuous Random Variables 66
3.13. Subroutines 72
3.14. Problems 72

Chapter 4. Discrete Probability Distributions... 76
4.1. Introduction 76
4.2. Discrete Uniform Distribution.. 76
4.3. Relationship between binomial, multinomial, hypergeometric, and multivariate hypergeometric PDFs 77
4.4. Binomial Distribution 78
4.5. Multinomial Distribution 84
4.6. Hypergeometric Distribution 87
4.7. Multivariate Hypergeometric Distribution 91
4.8. Negative Binomial Distribution.. 92
4.9. Geometric Distribution 95
4.10. Poisson Distribution 98
4.11. Subroutines 101
4.12. Problems 104

Chapter 5. Continuous Probability Distributions 106
5.1. Introduction 106
5.2. Continuous Uniform Distribution 106
5.3. Normal Distribution 110
5.4. Student's t Distribution 115
5.5. Gamma Distribution 118
5.6. Exponential Distribution 121
5.7. Chi-Squared Distribution 124
5.8. F distribution 126
5.9. Functions of Random Variables 128
5.10. Problems 131

Chapter 6. Sampling and Estimation 134
6.1. Introduction 134
6.2. Statistics 134
6.3. Sampling Distributions 136
6.4. Confidence Intervals 145
6.5. Problems 155

Appendix I. Review Handout for Discrete Probability Distribution Functions 158
Appendix II. Review Handout for Continuous Probability Density Functions 159
Appendix III. Review Handout for Statistics 160
Appendix IV. Summary of MATLAB Statistics Commands & References 162
Appendix V. Derivation of the fact that the distribution of the mean is normal 166
References 170

List of Subroutines

Chapter 1. Probability
    Code 1.1. Permutations via a Naïve Implementation (perm_naive)   23
    Code 1.2. Permutations with cancellations (perm)   23
    Code 1.3. Combinations with cancellations (comb)   23

Chapter 3. Expectations
    Code 3.1. Variance as a function of truncation (vartrunc)   72

Chapter 4. Discrete Probability Distributions
    Code 4.1. Binomial probability distribution (binomial)   101
    Code 4.2. Cumulative Binomial probability distribution (binocumu)   101
    Code 4.3. Arbitrary Ranges of Binomial probability distribution (binoprob)   101
    Code 4.4. Multinomial probability distribution (multinomial)   102
    Code 4.5. Hypergeometric probability distribution (hypergeo)   102
    Code 4.6. Multivariate Hypergeometric probability distribution (multihypergeo)   102
    Code 4.7. Negative Binomial probability distribution (negbinomial)   103
    Code 4.8. Geometric probability distribution (geo)   103
    Code 4.9. Arbitrary Ranges of the Geometric probability distribution (geoprob)   103
    Code 4.10. Poisson probability distribution (poisson)   103
    Code 4.11. Arbitrary Ranges of the Poisson probability distribution (poisprob)   103

Chapter 1. Probability

1.1. Introduction

In our quest to describe the properties of materials with statistical accuracy, we must first learn the language of statistics. Statistics itself is an applied field built upon the theory of probability, a mathematical discipline. Therefore, it is essential that we familiarize ourselves with the most basic premises of probability theory.

This chapter introduces the vocabulary of probability, counting rules, and rules for the probabilities of intersections, unions and conditional relationships.

1.2. Vocabulary

In this section, we introduce the vocabulary required for talking about probability.

Set

A set is a collection of objects or outcomes. Braces denote a set. For example, the set of letters in the alphabet is

$$S = \{a,b,c,d,e,f,g,h,i,j,k,l,m,n,o,p,q,r,s,t,u,v,w,x,y,z\} \tag{1.1}$$

The set of integers greater than 1 and less than 5 is

$$S = \{2,3,4\} \tag{1.2}$$

Element

An element is one member of a set. For example, 'a' is an element of the set, S, defined in equation (1.1).

Rule

The elements of a set can frequently be described by a rule. For example, we can rewrite the set of integers greater than 1 and less than 5 as

$$S = \{n \in I | 1 < n < 5\} \tag{1.3}$$

The symbol, $\in$, means 'is an element of' and the pipe reads as 'such that", so this statement should be read as the set including all n that are elements of the set of integer numbers (I), such that n is greater than 1 and less than 5.

Some sets must be written with a rule because they have a large or infinite number of elements that cannot be listed explicitly. For example, the set including all x that are elements of the set of real numbers ($\Re$), such that x is greater than 1 and less than 5 is written as

$$S = \{x \in \Re | 1 < x < 5\} \tag{1.4}$$

Another example of a rule is the set of ordered pairs (x,y) such that the satisfy a specific equation.

$$S = \{x, y | x^2 + y^2 \leq 4\} \tag{1.5}$$

Here the number of real solutions to the equation is infinite. Also an element of this example is an ordered pair.

Subset

A subset is part of a larger set. For example, the set of vowels are a subset of the set of letters, given in equation (1.1).

$$V = \{a, e, i, o, u\} \tag{1.6}$$

Null set

The null set is a set with no elements in it. The null set is written,

$$N = \{\emptyset\} \tag{1.7}$$

Complement

The complement of a subset, A, of a set, S, is defined as A' and includes all elements of S that are not in A. For example, the complement of the set of vowels, equation (1.6), from the set of letters, equation (1.1), is the set of consonants,

$$V' = \{b, c, d, f, g, h, j, k, l, m, n, p, q, r, s, t, v, w, x, y, z\} \tag{1.8}$$

The complement of the set, S, is the null set, N. The complement of the null set, N, is the entire set, S. The complement of a complement of a subset, A, is A, $(A')' = A$.

Sample Space

When we specifically apply this vocabulary to experiments, we say that the set of all possible outcomes of a statistical experiment is called the sample space, S. Several examples follow. When you flip a coin once, the sample space is heads or tails, $S=\{H,T\}$. When you toss a six-sided die, the sample space is a number from 1 to 6, $S=\{1,2,3,4,5,6\}$. When you flip two coins, the sample space is $S=\{HH,HT,TH,TT\}$.

When you measure the temperature of a furnace, the sample space includes all possible readings. For a digital read-out, accurate to the tenth of a degree and possessing four digits before the decimal point, the sample set could be written as $S=\{0.0,0.1,0.2...9999.7,9999.8,9999.9\}$. Of course, if the maximum safe, operating temperature of the furnace is 2000.0, then the probability of many of the elements in this sample space will be zero. Similarly, sample spaces can be defined for arbitrary experiments. If the data is recorded electronically, often it is output to 8 or 16 digits, many of which are not significant. Nevertheless, we often treat these samples spaces, while actually discrete (there are $10^{16}+1$ elements in a 16 digit number bound between 0.0 and 1.0 inclusively), such sample spaces are often treated as continuous.

Event

An event is a subset of a sample space. Consider the following example. In the experiment of flipping two coins, where the sample space is $S=\{HH,HT,TH,TT\}$, possible events include getting two heads, $B=\{HH\}$, getting the same result twice, $B=\{HH,TT\}$, getting a different result each time, $B=\{HT,TH\}$, or getting any result, $B=S$, or getting no result, $B=\{\varnothing\}$. (In a properly run experiment, the probability of getting the null set, should be zero.)

In the measurement of temperature, an event could be defined as all elements greater than the maximum safe operating temperature, $B = \{T|T > T_{max}\}$.

Intersection

The intersection of two events A and B, denoted by the symbol, $A \cap B$, is the event containing all elements in both A <u>and</u> B. The key operating word for the intersection is "and". The elements in the intersection are in A <u>and</u> B. For example, when we consider the example of flipping two coins,

If $B=\{HH,TT\}$ and $A=\{HH,HT,TH\}$, then $A \cap B = \{HH\}$.
If $B=\{HH,TT\}$ and $A=\{HT,TH\}$ then $A \cap B = \{\varnothing\}$.
If $B=\{HH,TT\}$ and $A=\{HH,TT\}$, then $A \cap B = A = B$.

As another example, consider testing the tensile strength of a material. An event A may be defined for samples that were exposed to a heat treatment (furnace temperature) of less than 1000 °C, $A = \{samples | T > 1000\}$. An event B may be defined for samples that possess satisfy the minimum requirements for the elastic modulus, E, greater than 70 GPa, $B = \{samples | E > 70\}$. The intersection of A and B is $A \cap B = \{samples | T > 1000 \text{ and } E > 70\}$. If this treatment

Probability - 5

temperature is sufficient to produce this tensile strength, the intersection of these two events will be large. If the treatment temperature is not sufficient to produce this tensile strength, the intersection may be small or empty.

Mutually exclusive, or disjoint

Two events A and B are mutually exclusive if there intersection is the null set, $A \cap B = \{\emptyset\}$, that is, if A and B have no common elements. For example: $A \cap A' = \{\emptyset\}$, that is, an event and its complement are by definition mutually exclusive events.

Union

The union of two events A and B, denoted by the symbol, $A \cup B$, is the event containing all elements in either A or B. The key operating word for the union is "or". The elements in the union are in A or B. For example, $A \cup A' = S$, that is, the union of an event and its complement is by definition mutually the sample space.

Venn Diagrams

Venn diagrams are a graphical way to express sets and events. The bounding box of the Venn Diagram contains the entire sample space, S. In Figure 1.1., we provide several examples of the use of Venn Diagrams in graphically interpreting various combinations of intersections, unions, and complements of sets. There are three events in this sample space, the circle "C", the triangle, "T", and the rectangle "R". In general the order or operations is important for combinations of intersection, unions and complements. Therefore the use of parentheses is required.

In Figure 1.2., we provide another example of a Venn Diagram. This one is used to define sustainable development. Sustainable development has economic, environmental and social constraints. Those practices that satisfy all three criteria are deemed to be sustainable. Note that in this Figure, the sample space is not

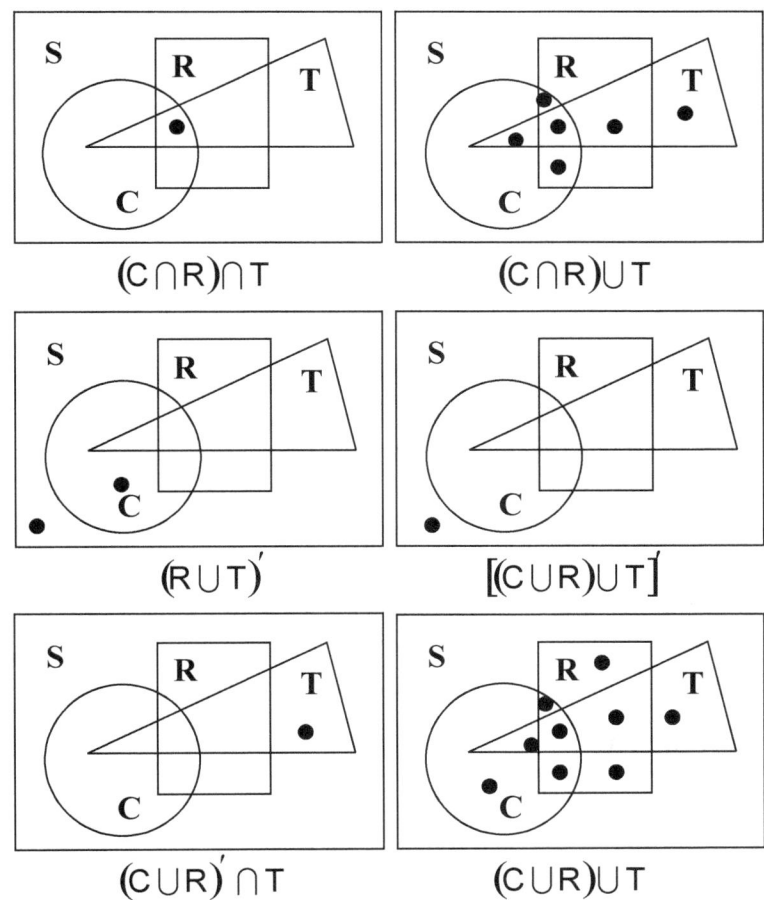

Figure 1.1. Visualizing probabilities through the use of Venn Diagrams.

explicitly defined. One could put the figure in a rectangle, defining the sample space S. Certainly there are some practices that meet none of the criteria.

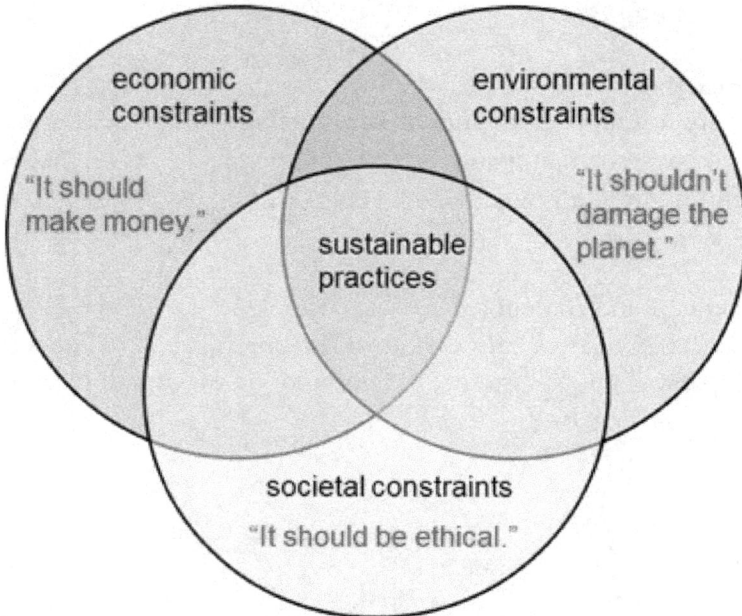

Figure 1.2. Defining sustainable development with a Venn Diagram.

1.3. Counting Rules

We need counting rules in probability because the probability of an event A is the ratio of the number of elements in A over the number of elements in the sample space S.

$$P(A) = \frac{\# of \text{ elements in } A}{\# of \text{ elements in } S} \tag{1.9}$$

Therefore, we need to know how to count the number of elements in A and S. We will study three counting rules:
1. generalized multiplication rule
2. permutations of distinct objects rule
3. combinations of distinct objects rule

Independence

Later in this text we will provide a formal definition of independence. However, we need a qualitative understanding of independence now. By independence, we mean that the result of the one experiment (or portion of an experiment) does not depend on the result of another experiment (or portion of an experiment). This is an important distinction and can be illustrated with the following simple example. Consider a bag containing 5 black marbles and 5 white marbles.

Consider an experiment where we randomly draw three marbles out of the bag, but we do so one marble at a time and replace the marble bag in the bag after each drawing. Clearly, the probability of getting a white marble on any single draw is always 0.5 via equation (1.9), because there are always 5 white and 5 black marbles in the bag.

$$P(W) = \frac{\#\,of\,\text{elements in}\,W}{\#\,of\,\text{elements in}\,S} = \frac{5}{10} = 0.5 \tag{1.10}$$

Thus sequential drawings of marbles with replacement are independent operations. The probability that we draw a white marble on the second draw is independent of the result of the first draw.

Now consider a second experiment in which we randomly draw three marbles from the same bag containing 5 black marbles and 5 white marbles, however this time we do so without replacing the marble. Therefore, the number of marbles changes each time we draw a marble. The probability of the first marble being white is still 0.5 as was the case above. However, the probability of drawing a white marble on the second draw depends on the result of the first draw. If the first draw was a white marble, then only 4 of the 9 remaining marbles are white and the probability of getting a white marble on the second draw given that we chose a white marble on the first draw is 4/9. On the other hand, if the first draw was a black marble, then 5 of the 9 remaining marbles are white and the probability of getting a white marble on the second draw given that we chose a black marble on the first draw is 5/9. Thus in this second experiment, each draw is not independent.

Generalized Multiplication rule

If an operation can be performed in n_1 ways, and if for each of these, a second independent operation can be performed in n_2 ways, and for each of the first two, a third independent operation can be performed in n_3 ways, and so forth, the sequence of k operations can be performed in

$$\#\,\text{of ways} = n_1 n_2 n_3 ... n_k \quad \text{for } k \text{ operations} \tag{1.11}$$

For example, if you consider the set of elements composed of one coin toss, followed by one six-sided die roll, followed by drawing from a hat containing m names. The number of elements in the set is 2*6*m. In this case it is clear that the operations are independent of each other. Flipping the coin does not change probability of each outcome on a die.

As another example, consider that you have a material with five compositions, which can be treated at one of four possible temperatures, for one of three durations of time. How many different outcomes are there? The answer, via the generalized multiplication rule, is that the
$\#\,\text{of ways} = n_1 n_2 n_3 = 5 \cdot 4 \cdot 3 = 60$.

Permutations

A permutation is an arrangement of all or part of a set of objects. A permutation is a grouping of elements arranged in a particular way. For example, how many ways can you order the letters A, B & C. The six permutations are ABC, ACB, BAC, BCA, CAB and CBA. All sequences contain the same letters but in different orders. The key concept in permutations is that 'order matters'.

The number of permutations of n distinct objects is $n!$ That is read as "n factorial".

$$n! = n(n-1)(n-2)(n-3)\ldots 3 \cdot 2 \cdot 1 \qquad (1.12)$$

The factorial only applies to non-negative integers. By definition, the factorial of zero is 1,

$$0! \equiv 1 \qquad (1.13)$$

The number of permutations of n distinct objects taken r at a time, where $r \leq n$, is

$$_nP_r = \frac{n!}{(n-r)!} \qquad (1.14)$$

For the example above, where we ordered three letters, $n = 3$ and $r = 3$, so the result is

$$_3P_3 = \frac{3!}{(3-3)!} = \frac{3!}{0!} = \frac{3 \cdot 2 \cdot 1}{1} = 6 \qquad (1.15)$$

When is the formula applicable? This formula applies when the order of a result is important and the objects are distinct. If the order doesn't matter, then there is only one way to take three letters from a set of three letters, namely a set that contains A, B and C. We shall discuss shortly how to count the number of ways when order doesn't matter. If the objects are not distinct, the permutation formula again does not apply. For example, if our three letters are A, B & A, even if order matters, we have only 3 arrangements, AAB, ABA and BAA. The number of arrangements are reduced because two of the elements were indistinguishable. We shall discuss shortly how to count the number of ways for indistinct objects.

As another example, we can ask how many ways can a group schedule 3 different meetings on any of five possible dates? The answer is $_5P_3 = 60$. How did we know to use the equation for permutations? The key tip-off was the word "different". This means the meetings are distinguishable and order matters.

Quick Calculation of Permutations by hand

When n becomes large but r is small, it can be difficult to compute the permutation of $_nP_r$. Consider the case where $n=200$ and $r=2$. Then

$$_nP_r = \frac{200!}{(200-2)!} = \frac{200!}{198!} \qquad (1.16)$$

Our calculators cannot compute the factorial of 200 or 198. The numbers are too large. However, we can still obtain the number of permutations, if we consider that

$$200! = 200 \cdot 199 \cdot 198! \qquad (1.17)$$

Then we have

$$_nP_r = \frac{200!}{(200-2)!} = \frac{200 \cdot 199 \cdot 198!}{198!} = 200 \cdot 199 = 39800 \qquad (1.18)$$

In general, we have

$$_nP_r = \frac{n!}{(n-r)!} = \frac{n \cdot (n-1) \cdot (n-2) \ldots (n-r+1)(n-r)!}{(n-r)!}$$
$$= n \cdot (n-1) \cdot (n-2) \ldots (n-r+1) = \prod_{i=n-r+1}^{n} i \qquad (1.19)$$

Codes for both the naïve and better implementations of permutations are provided in the Subroutines section of this chapter. These codes are written in a way that they exactly correspond to the theory presented above. Moreover, they are written in a way that they can be understood by students with a limited knowledge of programming.

Combinations

A combination is a grouping of elements without regard to order. The number of combinations of n distinct objects taken r at a time, where $r \leq n$, is

$$\binom{n}{r} = \frac{n!}{r!(n-r)!} \qquad (1.20)$$

The key for combinations is that order doesn't matter. In our example where we had three letters A, B, C and we chose three of them ($n=3$, $r=3$), we saw there were 6 permutations. From equation (1.20), we see that there is only one combination.

$$\binom{3}{3} = \frac{3!}{3!(3-3)!} = \frac{3!}{3!0!} = 1 \qquad (1.21)$$

In other words, there is only one way to take three objects from a pool of three objects if order doesn't matter, namely you take all of them. Again, this formula assumes that all of the objects are distinct.

Quick Calculation of Combinations by hand

When n becomes large but r is small, it can be difficult to compute the combination, $\binom{n}{r}$. The same cancellation trick that was used for permutations can again be used for combinations. Consider the case where $n=200$ and $r=2$. Then

$$\binom{n}{r} = \frac{n!}{r!(n-r)!} = \frac{200!}{2!(200-2)!} = \frac{200!}{2! \cdot 198!} = \frac{200 \cdot 199 \cdot 198!}{2! \cdot 198!} = \frac{200 \cdot 199}{2!} = 19900 \quad (1.22)$$

In general, we have

$$\binom{n}{r} = \frac{n!}{r!(n-r)!} = \frac{n!}{x_{big}! x_{lit}!} = \frac{n \cdot (n-1) \cdot (n-2) \ldots (x_{big}+1) x_{big}!}{x_{big}! x_{lit}!}$$
$$= \frac{n \cdot (n-1) \cdot (n-2) \ldots (x_{big}+1)}{x_{lit}!} \quad (1.23)$$

Here, we need to cancel either $r!$ or $(n-r)!$, whichever is larger. To make this clear we introduced two new variables where x_{big} is the larger of r and $n-r$ and x_{big} is the smaller. A code for this implementation of the combination is provided in the Subroutines section of this chapter.

In order to emphasize the difference between permutations and combinations, let us look at a few examples. We have already seen that considering the set of letters, A, B & C, for n = 3 and r = 3, that there are six permutations and one combination. If we only select two of the three letters, then we have $_3P_2$ =6 permutations. They are {AB, AC, BA, BC, CA, CB}. We have $\binom{3}{2} = 3$ combinations. They are {AB, BC, AC}.

As another example, consider a polymer film to be applied to a surface. The film is a composite of three layers, one to provide oxygen impermeability (O), one to provide water impermeability (W) and one to provide mechanical strength (S). If the order of the films is not important, then this calls for the combination rule, $\binom{3}{3} = 1$. There is only one way to lay down all three films. If the order of the films is important, then this is a permutation problem, $_3P_3 = 6$. There are six ways to lay down three films, {OWS, OSW, WOS, WSO, SOW, SWO}.

Probability - 11

Permutations of indistinct objects

In some cases, you have both distinct and indistinct objects. In this case, you must combine the permutation and combination formulae. The number of distinct permutations of n things where there are n_1 of one kind, n_2 of the second kind, up to n_k of the kth kind is:

$$permutations\_of\_indistinct\_objects = \frac{n!}{n_1! n_2! ... n_k!} \qquad (1.24)$$

For example, how many ways can you arrange all elements of the following set {A,A,A,B,B,C}?

$$\frac{6!}{3!2!1!} = \frac{720}{1} * \frac{1}{12} = 60 \qquad (1.25)$$

Consider an example where we have a three letter passcode that can be composed of any of the 26 letters in the alphabet.

Q: How many combinations are there if each letter is used no more than once?
A: This problem requires the permutation formula with $n = 26$ and $r = 3$, because each object is distinct.

$$_{26}P_3 = \frac{26!}{(26-3)!} = \frac{26!}{23!} = \frac{26 \cdot 25 \cdot 24 \cdot 23!}{23!} = 26 \cdot 25 \cdot 24 = 15,600 \qquad (1.26)$$

Q: How many combinations are there if each letter can be reused?
A: This problem requires the general multiplication rule because the choice of each letter is independent.

$$\# of\ ways = 26 \cdot 26 \cdot 26 = 17,576 \qquad (1.27)$$

1.4. Probability

The probability of an event A is the sum of the weights of all sample points in A. By weights, we mean the number of times a particular outcome is represented. The weight of A in the set {A,A,B} is 2. The weight of B is one.

The probability of an event A must satisfy the following conditions. First, the probability is always bounded between zero and one.

$$0 \le P(A) \le 1 \qquad (1.28)$$

There is no such thing as a negative probability. There is also no probability greater than 100%. Second, the probability of the null set is zero.

$$P(\emptyset) = 0 \tag{1.29}$$

This is another way of saying that the experiment will result in some outcome. Third, the probability of getting something in the sample space is one.

$$P(S) = 1 \tag{1.30}$$

Since the sample space contains all possible outcomes, the outcome has to lie within the sample space.

Consider the following example. On an ordinary six-sided die, any number is equally likely to turn up. The probability of getting any particular number is 1/6, which is between 0 and 1. The probability that you don't get any result (the null set) when you roll the die is zero. The problem that you get something in the set $S=\{1,2,3,4,5,6\}$ is one.

Now consider an unbalanced six-sided die that is weighted to preferentially yield 6. For example, instead of yielding each number between 1 and 6 1/6 of the time, the die yields 6 half the time, and the rest of the numbers 1/10 of the time. All of the probabilities are still bounded between 0 and 1. The sum of the weights $5*(0.1) + 0.5 = 1$. Therefore, the probability of getting a 6 is $P(6) = 0.5$. The probability of any other number, such as 1, is $P(1) = 0.1$.

Union

The probability of the union of two events A and B is given by

$$P(A \cup B) = P(A) + P(B) - P(A \cap B) \tag{1.31}$$

For example, the sample space consists of the letter number pairs, $S=\{A1, A2, B1\}$. The probability of getting a pair with an A or a pair with a 1 is

$$\begin{aligned} &P(A \cup 1) = P(A) + P(1) - P(A \cap 1) \\ &P(A) = 2/3 \\ &P(1) = 2/3 \\ &P(A \cap 1) = 1/3 \\ &P(A \cup 1) = 2/3 + 2/3 - 1/3 = 1 \end{aligned} \tag{1.32}$$

If A and B are mutually exclusive, then their union is the sum of their probabilities (because their intersection is zero).

$$P(A \cup B) = P(A) + P(B) \tag{1.33}$$

If many events, $A_1 A_2 \ldots A_n$ are mutually exclusive, then their union is the sum of their probabilities.

$$P(A_1 \cup A_2 \cup ... \cup A_n) = \sum_{i=1}^{n} P(A_i) \qquad (1.34)$$

If many events, A_1 A_2....A_n, are mutually exclusive and include all of the sample space,

$$P(A_1 \cup A_2 \cup ... \cup A_n) = \sum_{i=1}^{n} P(A_i) = P(S) = 1 \qquad (1.35)$$

The union of three events is given by

$$P(A \cup B \cup C) = P(A) + P(B) + P(C) - P(A \cap B) - P(A \cap C) - P(B \cap C) \\ + P(A \cap B \cap C) \qquad (1.36)$$

For example, consider the set of ten objects S = {cat, dog, wolf, tiger, oak, elm, maple, opal, ruby, pearl}.

Q: What is the probability that you would randomly select a word that is (A) an animal OR that (B) has 4 letters OR that (C) starts with a vowel.)

A: There are two methods of solution. First, you can pick these out by hand. There are 4 animals, 2 trees that start with vowels, and 2 minerals that have 4 letters. Therefore the probability is 0.8 or 80%. The other method of solution is to use the equation given above to find the probability of the union.

$$\begin{aligned} &P(A) = 4/10; \quad P(B) = 3/10; \; P(C) = 3/10; \\ &P(A \cap B) = 1/10; \quad P(A \cap C) = 0; \; P(B \cap C) = 1/10; \; P(A \cap B \cap C) = 0; \\ &P(A \cup B \cup C) = 4/10 + 3/10 + 3/10 - 1/10 - 0 - 1/10 + 0 = 8/10 \end{aligned} \qquad (1.37)$$

Conditional Probability

The probability of an event B occurring when it is know that some event A has already occurred is called a conditional probability, $P(B|A)$, and is read, "the probability of B given A", and is defined by:

$$P(B \mid A) = \frac{P(A \cap B)}{P(A)} \qquad \text{for } P(A) > 0 \qquad (1.38)$$

Using the above example of ten words, what is the probability that we choose a (B) three letter word given that we know that we have chosen an (A) animal.

$$P(B \mid A) = \frac{P(A \cap B)}{P(A)} = \frac{2/10}{4/10} = 1/2 \qquad (1.39)$$

The conditional probability of B given A is different than the conditional probability of A given B,

$$P(A \mid B) = \frac{P(A \cap B)}{P(B)} \qquad \text{for } P(B) > 0 \tag{1.40}$$

Equations (1.38) and (1.40) can be rearranged in terms of the intersection

$$P(A \cap B) = P(A \mid B)P(B) = P(B \mid A)P(A) \tag{1.41}$$

If it possible to compute the probability of the intersection of A and B, given information about the conditional probability and the probability of A or B. This equation also provides a relationship between the conditional probability of A given B and the conditional probability of B given A.

The rule for the intersection can be extended to more than two events. If in an experiment, $A_1, A_2, A_3 ... A_k$ can occur then:

$$P(A_1 \cap A_2 \cap ... \cap A_k) = P(A_1)P(A_2 \mid A_1)P(A_3 \mid A_1 \cap A_2) .. P(A_k \mid A_1 \cap A_2 \cap ... \cap A_{k-1}) \tag{1.42}$$

This formula stems from the repeated application of the conditional probability rule.

Independence/Multiplicative Rules

If two events are independent, then the outcome of the first experiment does not impact the outcome of the second experiment. Thus the probability of B happening should not be affected by the probability that A happened. In other words, two events are independent if and only if

$$P(B \mid A) = P(B) \quad \text{and} \quad P(A \mid B) = P(A) \quad \text{if } A \text{ and } B \text{ are independent} \tag{1.43}$$

Substituting eqns (1.43) into equation (1.41) we have that two events are independent if and only if

$$P(A \cap B) = P(A) * P(B) \quad \text{if } A \text{ and } B \text{ are independent} \tag{1.44}$$

This expression can be used as a way to test for independence, if the probability of the intersection is known. For multiple events, we have that the events are independent if and only if

$$P(A_1 \cap A_2 \cap ... \cap A_k) = \prod_{i=1}^{k} P(A_i) \quad \text{if all } A_i \text{ are independent} \tag{1.45}$$

Example Problems for Probability

There are three examples given below. In each example, let it be clear that we use three and only three rules! We use the rules for the union, the conditional probability, and the intersection. To restate these rules, we have for two elements:

Union: $P(A \cup B) = P(A) + P(B) - P(A \cap B)$ (1.31)

Conditional: $P(B|A) = \dfrac{P(A \cap B)}{P(A)}$ for $P(A) > 0$ (1.38)

Intersection: $P(A \cap B) = P(A)P(B|A) = P(B)P(A|B)$ (1.41)

Example 1.1: You flip a coin twice. What is the probability of getting heads on the second flip (B) given that you got heads on the first flip (A)? Are the events independent?

Solution:
The probability of getting a head on the first flip is 0.5 The probability of getting a head on the second flip is 0.5. The intersection of A and B from the set {HH, HT, TH, TT} is 0.25.
The conditional probability is thus

$$P(B|A) = \dfrac{P(A \cap B)}{P(A)} = \dfrac{0.25}{0.5} = 0.5 \quad (1.46)$$

We see that $P(B|A) = P(B) = 0.5$ so each flip of the coin is independent.

Example 1.2: You have a bag with 3 lima beans and 2 pinto beans in it. You draw 2 beans from it randomly without replacement. What is the probability that you draw a lima bean (B) given that you already drew a lima bean on the first draw (A)? Are the events independent?

Solution:
The easiest way to solve this problem is to list all the possible outcomes.
The probability for drawing a lima bean the first time (event A) is 3/5 = 0.6
The probability for drawing a pinto bean the first time is 2/5 = 0.4
If we draw a lima bean the first time, then there are 2 lima beans and 2 pinto beans. In that case the probability for drawing a lima bean the second time is 2/4 = 0.5 and the probability for drawing a pinto bean the second time is 2/4 = 0.5.
If we draw a pinto bean the first time, then there are 3 lima beans and 1 pinto beans. In that case the probability for drawing a lima bean the second time is 3/4 = 0.75 and the probability for drawing a pinto bean the second time is 1/4 = 0.25.
So we have four possible outcomes {LL, LP, PL, PP}.
The probability of each outcome is given by the product of the probabilities of the events in that outcome. (This is the generalized multiplicative rule.)
The probability of LL is 0.6*0.5 = 0.3.
The probability of LP is 0.6*0.5 = 0.3.
The probability of PL is 0.4*0.75 = 0.3.
The probability of PP is 0.4*0.25 = 0.1.

The individual probabilities then for {LL, LP, PL, PP} are {0.3, 0.3, 0.3, 0.1}. With these figures, we can write:

Now event A includes any outcome with L in the first draw, LL and LP.

$$P(A) = P(LL) + P(LP) = 0.3 + 0.3 = 0.6 \qquad (1.47)$$

Event B includes any outcome with L in the first draw, LL and PL. The sum of those two probabilities is 0.3+0.3 = 0.6 so

$$P(B) = P(LL) + P(PL) = 0.3 + 0.3 = 0.6 \qquad (1.48)$$

The intersection of A and B includes the outcome LL

$$P(A \cap B) = P(LL) = 0.3 \qquad (1.49)$$

Given this information, we have that the conditional probability of B given A, (or the probability that we draw a lima bean on the second draw, given that we drew a lima bean on the first draw) is

$$P(B \mid A) = \frac{P(A \cap B)}{P(A)} = \frac{0.3}{0.6} = 0.5 \qquad (1.50)$$

Now, to check for independence, we can evaluate equation (1.43)

$$P(B \mid A) = 0.5 \neq P(B) = 0.6 \qquad (1.51)$$

Therefore, the two experiments are not independent. We can equivalently check for independence by examining the intersection of A and B from equation (1.44).

$$P(A \cap B) = 0.3 \neq P(A) * P(B) = 0.6 * 0.6 = 0.36 \qquad (1.52)$$

This also says events A and B are not independent. In general, there is no need to make both checks. They will always return the same result. Use the one for which the information is more readily available.

Example 1.3.
In sampling a population for the presence of a disease, the population is of two types: Infected (I) and Uninfected (U). The results of the test are of two types: Positive (P) and Negative (N). In rare disease detection, a high probability for detecting a disease can still lead to more false positives than true positives. Consider a case where a disease infects 1 out of every 100,000 individuals. The probability for a positive test result given that the subject is infected is 0.99. (The test can accurately identify an infected individual 99% of the time.) The probability for a

negative test result given that the subject is uninfected is 0.999. (The test can accurately identify that an uninfected individual 99.9% of the time.)

We shall answer the following questions.

(1) For testing a single person, define the complete sample space.

(2) What is the probability of a false negative test result (a negative test result given that the subject is infected)?

(3) What is the probability of being uninfected AND having a negative test result?

(4) What is the probability of testing positive?

(5) Determine rigorously whether testing positive and having the disease are independent.

(6) Determine the percentage of people who test positive who are really uninfected.

(7) In a population of 250 million, with the infection rate given, how many people would you expect to be (a) Infected-test Positive, (b) Infected-test Negative, (c) Uninfected-test Positive, (d) Uninfected-test negative.

Solution:

We are given the following information.

$$P(I) = 10^{-5} \tag{1.53}$$

$$P(N|U) = 0.999 \tag{1.54}$$

$$P(P|I) = 0.99 \tag{1.55}$$

(1) For testing a single person, define the complete sample space.

The sample space is $S = \{IP, IN, UP, UN\}$ where I = Infected, U=Uninfected, P=positive test result, N=negative test result. The Venn Diagram looks like this:

Infected ∩ Positive	Infected ∩ Negative
Uninfected ∩ Positive	UnInfected ∩ Negative

When you have a simple sample space like this, you can see some additional constraints on the system, in addition to the union, conditional, and intersection rules. You will need some of these additional constraints to solve the problems below.

For example, if a person tests positive, they are either infected or uninfected. Therefore, using the union rule we have:

$$P(P) = P[(P \cap I) \cup (P \cap U)] = P(P \cap I) + P(P \cap U) - P[(P \cap I) \cap (P \cap U)] \tag{1.56}$$

There is no intersection between being infected and uninfected, therefore:

$$P(P) = P(P \cap I) + P(P \cap U) \tag{1.57}$$

We can write three other analogous constraints based on the union rule,

$$P(N) = P(N \cap I) + P(N \cap U) \tag{1.58}$$
$$P(U) = P(U \cap P) + P(U \cap N) \tag{1.59}$$
$$P(I) = P(I \cap P) + P(I \cap N) \tag{1.60}$$

Also consider that the probability of being infected given a person is positive plus the probability of being uninfected given a person is positive is 1. A person is either infected or uninfected, regardless of whether they tested positive or negative. We can write this as.

$$P(I \mid P) + P(U \mid P) = \frac{P(I \cap P)}{P(P)} + \frac{P(U \cap P)}{P(P)} = \frac{P(P)}{P(P)} = 1 \tag{1.61}$$

Here we used the conditional probability rule and the union rule. This leads to the following constraint and three analogous constraints,

$$P(I \mid P) + P(U \mid P) = 1 \tag{1.61}$$
$$P(I \mid N) + P(U \mid N) = 1 \tag{1.62}$$
$$P(P \mid U) + P(N \mid U) = 1 \tag{1.63}$$
$$P(P \mid I) + P(N \mid I) = 1 \tag{1.64}$$

In solving the problems, below, remember we have this group of rules. There are many ways to solve some of the problems. We just go looking for the one that seems easiest.

(2) What is the probability of a false negative test result (a negative test result given that the subject is infected)?

We want: $P(N|I) = \dfrac{P(N \cap I)}{P(I)}$ so we need the two factors on the right hand side. We have been given the denominator. In order to find the numerator, we must use the other given:

$$P(P|I) = \frac{P(P \cap I)}{P(I)} = 0.99 \tag{1.65}$$

which rearranges for the intersection of P and I

$$P(P \cap I) = P(I) \cdot P(P|I) = (10^{-5})(0.99) = 0.99 \cdot 10^{-5} \tag{1.66}$$

We must realize that the probability of I is the union of IP and IN groups, equation (1.60). So using the definition of the Union, we have:

$$P(I) = P(I \cap P) + P(I \cap N) \tag{1.60}$$

Rearranging yields

$$P(I \cap N) = P(I) - P(I \cap P) = (10^{-5}) - 0.99 \cdot 10^{-5} = 10^{-7} \tag{1.67}$$

Then we can plug into our original equation:

$$P(N|I) = \frac{P(N \cap I)}{P(I)} = \frac{10^{-7}}{10^{-5}} = 0.01 \tag{1.68}$$

OR, an alternative solution, relies on us recognizing

$$P(P|I) + P(N|I) = 1 \tag{1.64}$$

and rearranging

$$P(N|I) = 1 - P(P|I) = 1 - 0.99 = 0.01 \tag{1.69}$$

(3) What is the probability of being uninfected AND having a negative test result?

We want $P(N \cap U)$. We can obtain this from either:

(a) the UNION RULE:

$$P(N) = P[(N \cap I) \cup (N \cap U)]$$
$$P(N) = P(N \cap I) + P(N \cap U) - P[(N \cap I) \cap (N \cap U)]$$
$$P[(N \cap I) \cap (N \cap U)] = 0$$
$$P(N) = P(N \cap I) + P(N \cap U)$$
$$P(N \cap U) = P(N) - P(N \cap I)$$

but we don't know $P(N \cap I)$ and we don't know $P(N)$, so this doesn't seem immediately helpful.

or (b) the conditional probability rule:

$$P(U|N) = \frac{P(N \cap U)}{P(N)}$$

$$P(N \cap U) = P(N) \cdot P(U|N)$$

but we don't know $P(U|N)$ and we don't know $P(N)$, so again this doesn't seem immediately helpful.

or (c) the conditional probability rule:

$$P(N|U) = \frac{P(N \cap U)}{P(U)} = 0.999$$

$$P(N \cap U) = P(U) \cdot P(N|U) = P(U) \cdot 0.999$$

I like choice (c) because we are given $P(N|U) = 0.999$ and we know

$$P(U) = 1 - P(I) = 1 - 10^{-5} = 0.99999 \text{ so}$$

$$P(N \cap U) = P(U) \cdot P(N|U) = (0.99999)(0.999) = 0.99899001 \tag{1.70}$$

(4) What is the probability of testing positive?

We want $P(P)$. We can find $P(P)$ by either:

(a) the fact that the sum of the probabilities must be one

$$P(P) + P(N) = 1$$

$$P(P) = 1 - P(N)$$

but we don't know $P(N)$.

or (b) the conditional probability distribution:

$$P(I|P) = \frac{P(P \cap I)}{P(P)} \text{ but we don't know } P(I|P).$$

or (c) the conditional probability distribution:

$$P(U|P) = \frac{P(P \cap U)}{P(P)} \text{ but we don't know } P(U|P).$$

or (d) the sum of the probabilities must be one and a different conditional probability:

Probability - 21

$$P(P) = 1 - P(N)$$
$$P(U|N) = \frac{P(N \cap U)}{P(N)}$$
$$P(P) = 1 - P(N) = 1 - \frac{P(N \cap U)}{P(U|N)} \text{ but we don't know } P(U|N).$$

or (e) the sum of the probabilities must be one and a different conditional probability:

$$P(P) = 1 - P(N)$$
$$P(I|N) = \frac{P(N \cap I)}{P(N)}$$
$$P(P) = 1 - P(N) = 1 - \frac{P(N \cap I)}{P(I|N)} \text{ but we don't know } P(I|N).$$

or (f) the union rule:

$$P(P) = P[(P \cap I) \cup (P \cap U)]$$
$$P(P) = P(P \cap I) + P(P \cap U) - P[(P \cap I) \cap (P \cap U)]$$
$$P[(P \cap I) \cap (P \cap U)] = 0$$
$$P(P) = P(P \cap I) + P(P \cap U)$$

Combine this expression with conditional probabilities that we do know:

$$P(P) = P(I) * P(P|I) + P(U) * P(P|U)$$

I like choice (f).

$$P(P) = 10^{-5} \cdot 0.99 + 0.99999 \cdot P(P|U)$$

We can get the last factor by considering

$$P(P|U) + P(N|U) = 1 \tag{1.63}$$

$$P(P|U) = 1 - P(N|U) = 1 - 0.999 = 0.001$$

so

$$P(P) = 10^{-5} \cdot 0.99 + 0.99999 \cdot 0.001 = 0.00100989 \tag{1.71}$$

(5) Determine rigorously whether testing positive and having the disease are independent.

If $P(P)$ and $P(I)$ are independent, then

$$P(P \cap I) = P(P) \cdot P(I).$$

$$0.99 \cdot 10^{-5} = 0.00100989 \cdot 10^{-5}$$

Testing positive and being infected are not independent. Thank goodness. The entire point of the test is to identify infected individuals.

(6) Determine the percentage of people who test positive but who are really uninfected.

We want $\dfrac{P(P \cap U)}{P(P)}$.

$$\frac{P(P \cap U)}{P(P)} = \frac{0.99999 \cdot 10^{-3}}{0.00100989} = 0.990196952 = 99\%$$

<u>Despite the high accuracy of the test 99% of those people who test positive are actually uninfected.</u>

(7) In a population of 250 million, with the infection rate given, how many people would you expect to be (a) Infected-test Positive, (b) Infected-test Negative, (c) Uninfected-test Positive, (d) Uninfected-test negative.

From part (5) we know:

$$P(P \cap I) = 0.99 \cdot 10^{-5}$$
$$P(P \cap U) = 0.99999 \cdot 0.001 = 0.99999 \cdot 10^{-3}$$

From part (2) we know

$$P(N \cap I) = P(I) - P(P \cap I) = (10^{-5}) - 0.99 \cdot 10^{-5} = 10^{-7}$$

From part (3) we know that

$$P(N \cap U) = P(U) \cdot P(N|U) = (0.99999)(0.999) = 0.99899001$$

These four probabilities should sum to 1.0 and they do.

Out of 250 million people, the number who are infected and test positive are: 2475.
Out of 250 million people, the number who are infected and test negative are: 25.
Out of 250 million people, the number who are uninfected and test positive are: 249,997.5
Out of 250 million people, the number who are uninfected and test negative are: 249,747,500.

1.5 Subroutines

Code 1.1. Permutations via a Naïve Implementation (perm_naive.m)

This simple code, perm_naive.m, illustrates how one would numerically compute a permutation. It doesn't use the cancellation trick shown above. It computes the factorial of n, then computes the factorial of $(n-x)$, then returns the quotient. This code won't work for large numbers

```
function f = perm_naive(n,x)
fac1 = 1.0;
if (n > 1)
    for i = n:-1:2
        fac1 = fac1*i;
    end
end
fac2 = 1.0;
if (n-x > 1)
    for i = (n-x):-1:2
        fac2 = fac2*i;
    end
end
f = fac1/fac2;
```

Code 1.2. Permutations with cancellations (perm.m)

This code, perm.m, computes permutations, using the cancellation trick.

```
function f = perm(n,x)
f = 1.0;
if (n > 1 & n >=x)
    for i = n:-1:(n-x+1)
        f = f*i;
    end
end
```

Code 1.3. Combinations (with cancellations) (comb.m)

This code, comb.m, illustrates how one would numerically compute a combination. It invokes the cancellation trick, so it must first determine the larger factorial in the denominator.

```
function f = comb(n,x)
a = n-x;
if (a > x)
```

```
        xbig = a;
        xlit = x;
    else
        xbig = x;
        xlit = a;
    end
    if (n > 1 & n >=xbig)
        fnum = 1.0;
        for i = n:-1:(xbig+1)
            fnum = fnum*i;
        end
        fden = 1.0;
        for i = xlit:-1:2
            fden = fden*i;
        end
    end
    f = fnum/fden;
```

1.6. Problems

Problem 1.1.
Consider sputtering an film with 8 layers that are designated A-rich, A-lean, B-rich, B-lean, C-rich, C-lean, D-rich and D-lean. In how many different ways can the layers be put down
  (a) with no restrictions?
  (b) if the corresponding rich and lean layers must be adjacent?
  (c) if all the rich layers are sputtered first, followed by all the lean layers?

Problem 1.2.
  (a) A preliminary engineering design involves three stages where different solvents are used to perform liquid-liquid extraction. If you are considering 5 different solvents, each of which will be used in no more than one extractor, how many different possible designs would you need to investigate?
  (b) Now make the assumption that it doesn't matter what order the three extractions are performed in. How many different possible designs would you need to investigate?

Problem 1.3.
A drug for the relief of arthritis can be purchased from 8 different manufacturers in liquid, tablet, or capsule form, all of which come in 100, 200 and 400 mg doses. How many different ways can a doctor prescribe the drug to a patient suffering from arthritis?

Problem 1.4.
A group of 500 college freshman are surveyed on the subject of fossil energy, climate change and sustainability. 210 believe sustainability is a legitimate global issue. 258 believe that anthropogenic climate change exists. 216 believe that fossil energy will be exhausted in their lifetime. 122 believe in both sustainability and anthropogenic climate change. 83 believe in both

the exhaustion of fossil energy and anthropogenic climate change. 97 believe in both the sustainability and the exhaustion of fossil energy. 52 believe all three statements.

(a) Find the probability of those who believe in sustainability but not in anthropogenic climate change.
(b) Find the probability of those who believe fossil fuel will be exhausted in their lifetimes and in anthropogenic climate change but not in sustainability.
(c) Find the probability of those who believe in neither sustainability nor that fossil energy will be exhausted.

Problem 1.5.
In sampling a population for the presence of a disease, the population is of two types: Infected and Uninfected. The results of the test are of two types: Positive and Negative. In rare disease detection, a high probability for detecting a disease can still lead to more false positives than true positives. Consider a case where a disease infects 1 out of every 100,000 individuals. The probability for a positive test result given that the subject is infected is 0.99. The probability for a negative test result given that the subject is uninfected is 0.999.

(a) What is the probability of being uninfected?
(b) What is the probability of being uninfected AND testing negative?
(c) What is the probability of being uninfected AND testing positive?
(d) What is the probability of testing positive?
(e) What is the probability of being uninfected given that the person tested positive? (This is the percentage of erroneous positive tests.)

(Make sure your answers are accurate to five significant figures.)

Chapter 2. Random Variables and Probability Distributions

2.1. Introduction

In the previous chapter, we introduced common topics of probability. In this chapter, we translate those concepts into a mathematical framework. We invoke algebra for discrete variables and calculus for continuous variables. Every topic in this chapter is presented twice, once for discrete variables and again for continuous variables. The analogy in the two cases should be apparent and should reinforce the common underlying concepts. There is, in the second half of the chapter, another duplication of concepts in which we show that the same process of translation from the language of probability to that of mathematics can be performed not only when we have a single variable of interest, but also when we have two variables. Again, high-lighting this analogy between single and joint probability distributions explicitly reveals the common underlying concepts.

2.2. Random Variables & Sample Spaces

We begin with the introduction of a necessary vocabulary.

Random variable
A random variable is a function that associates a number, integer or real, with each element in a sample space.

Discrete Sample Space
If a sample space contains a finite number of possibilities or an unending sequences with as many elements as there are whole numbers, it is called a discrete sample space.

Example 2.1.: You flip two coins. Y is a random variable that counts the number of heads. The possible results and the value of the random variable associated with each result are given in the following table.

result	y
HH	2
HT	1
TH	1
TT	0

This sample space is discrete because there are a finite number of possible outcomes.

Example 2.2.: You subject a box containing N devices to a test. Y is a random variable that counts the number of defective devices. The value of the random variable ranges from 0 (no defects) to N (all defects). This sample space is discrete because there are a finite number of possible outcomes.

Continuous Sample Space
If a sample space contains an infinite number of possibilities equal to the number of points on a line segment, it is called a continuous sample space.

Example 2.3.: You drive a car with five gallons of gas. Y is a random variable that represents the distance traveled. The possible results are infinite because even if the car averaged 20 miles per gallon, it could go 100.0 miles, 100.1, 100.01, 100.001, 100.0001 miles. The sample space is as infinite as real numbers.

2.3. Discrete Probability Distribution Functions (PDFs)

Probability distribution function (PDF)
The function, $f(x)$ is a probability distribution function of the discrete random variable x, if for each possible outcome a, the following three criteria are satisfied.

$$f(x) \geq 0$$
$$\sum_x f(x) = 1 \qquad (2.1)$$
$$P(x = a) = f(a)$$

The PDF is always non-negative. The PDF is normalized, meaning that the sum over all values of a discrete PDF is unity. The PDF evaluated at outcome a provides the probability of the occurrence of outcome a.

Example 2.4.: Eight devices are shipped to a retail outlet, 3 of which are defective. If a consumer purchases 2 computers, find the probability distribution for the number of defective devices bought by the consumer.

In order to solve this problem, first define the random variable and the range of the random variable. The random variable, x, is equal to the number of defective devices bought by the consumer. The random variable, x, can take on values of 0, 1, and 2. Those are the only number of defective devices the consumer can buy, given that they are only buying two devices.

The next step is to determine the size of the sample space. The number of ways that 2 can be taken from 8 without replacement is $\binom{8}{2} = 28$. We use the formula for combinations because the order of purchases does not matter. This is the total number of combinations of devices that the consumer can buy.

Third, the probability of a particular outcome is equal to the number of ways to get that outcome over the total number of ways:

$$f(x) = P(x = a) = \frac{\text{ways of getting a}}{\text{total ways}}$$

$$f(0) = P(x = 0) = \frac{\binom{3}{0}\binom{5}{2}}{28} = \frac{10}{28}$$

$$f(1) = P(x = 1) = \frac{\binom{3}{1}\binom{5}{1}}{28} = \frac{15}{28}$$

$$f(2) = P(x = 2) = \frac{\binom{3}{2}\binom{5}{0}}{28} = \frac{3}{28}$$

In each of these cases, we obtained the numerator, the number of ways of getting outcome a, by using the combination rule and the generalized multiplication rule. There are $\binom{3}{a}$ ways of choosing a defective devices from 3 defective devices. There are $\binom{5}{2-a}$ ways of choosing $(2-a)$ good devices from 5 good devices. We use the generalized multiplication rule to get the number of ways of getting both of these outcomes in the numerator. As a preview, we will come to discover that this probability distribution is called the hypergeometric distribution in Chapter 4.

So we have the PDF, f(x), defined for all possible values of x. We have solved the problem.

Note: If someone asked for the probability for getting 3 (or any number other than 0, 1, or 2) defective devices, then the probability is zero and f(3)= 0.

Testing a discrete PDF for legitimacy

If you are asked to determine if a given PDF is legitimate, you are required to verify the three criteria in equation (2.1).   Generally, the third criterion is given in the problem statement, so you only have to check the first two criteria.

The first criteria, $f(x) \geq 0$, can most easily be verified by plotting f(x) and showing that it is never negative. The second criteria, $\sum_x f(x) = 1$, can most easily be verified by direct summation of all *f(x)*.

Normalizing a discrete PDF

Discrete PDF's must satisfy $\sum_x f(x) = 1$. Sometimes, you have the functional form of the PDF and you simply need to force it to satisfy this criterion. In that case you need to normalize the PDF so that it sums to unity.

If $f^*(x)$ is an unnormalized PDF, then it can be normalized by the multiplication of a constant,

$$f(x) = cf^*(x) = \frac{1}{\sum_x f^*(x)} f^*(x) \qquad (2.2)$$

where that constant is the inverse of the sum of the unnormalized PDF.

Example 2.5.: Find the value of *c* that normalizes the following PDF.

$$f(x) = c\left(_4P_x\right) \text{ for } x = 0, 1, 2, 3, \& 4$$

To normalize, we sum the PDF over all values and set it to unity.

$$\sum_x f(x) = 1 = \sum_{x=0}^{4} f(x) = f(0) + f(1) + f(2) + f(3) + f(4)$$

$$\sum_x f(x) = 1 = \sum_{x=0}^{4} c\left(_4P_x\right) = c\sum_{x=0}^{4} \left(_4P_x\right) = c\left(_4P_0 + {_4P_1} + {_4P_2} + {_4P_3} + {_4P_4}\right) = c(1 + 4 + 12 + 24 + 24)$$

We then solve for simplify and solve for the normalization constant, *c*.

$$1 = c(65)$$

$$c = \frac{1}{65}$$

So the normalized PDF is

$$f(x) = \frac{1}{65}\left(_4P_x\right)$$

Discrete Cumulative Distribution Function (CDF)

The discrete cumulative distribution function (CDF), $F(x)$ of a discrete random variable X with the probability distribution, $f(x)$, is given by

$$F(a) = P(x \le a) = \sum_{x \le a} f(x) \quad \text{for} \quad -\infty < x < \infty \qquad (2.3)$$

The CDF is the probability that x is less than or equal to a.

Example 2.6.: In the above example, regarding the consumer purchasing devices, we can obtain the cumulative distribution directly:

$F(0) = f(0) = 10/28$,
$F(1) = f(0)+f(1) = 25/28$,
$F(2) = f(0)+f(1)+f(2) = 1$

Note: The cumulative distribution is always monotonically increasing, with x. The final value of the cumulative distribution is always unity, since the PDF is normalized.

Probability Histogram:

A probability histogram is a graphical representation of the distribution of a discrete random variable.

The histogram for the PDF and CDF for the Example 2.4. are given in Figure 2.1.

The histogram of the PDF provides a visual representation of the probability distribution, its most likely outcome and the shape of the distribution.

The histogram of the CDF provides a visual representation of the cumulative probability of an outcome. We observe that the CDF is monotonically increasing and ends at one, as it must since the PDF is normalized and sums to unity.

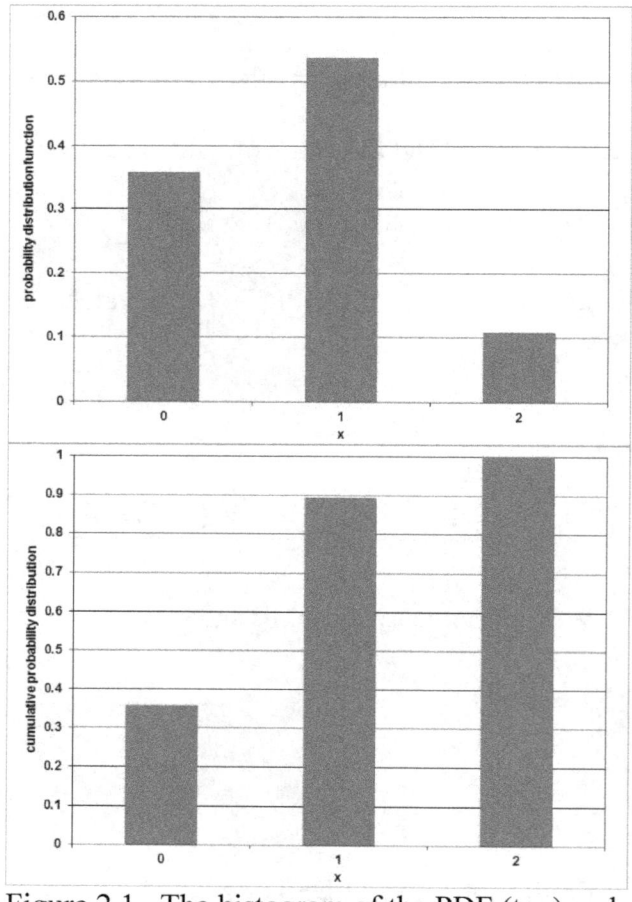

Figure 2.1. The histogram of the PDF (top) and CDF (bottom) for Example 2.4.

2.4. Continuous Probability Density Functions (PDFs)

Probability distribution functions of discrete random variables are called probability density functions when applied to continuous variables. Both have the same meaning and can be abbreviated commonly as PDF's. Probability density functions satisfy three criteria, which are analogous to those for discrete PDFs, namely

$$f(x) \geq 0 \quad \text{for all } x \in R$$

$$\int_{-\infty}^{\infty} f(x)dx = 1 \tag{2.4}$$

$$P(a < x < b) = \int_{a}^{b} f(x)dx$$

The probability of finding an exact point on a continuous random variable is zero,

$$P(x = a) = P(a \leq x \leq a) = \int_{a}^{a} f(x)dx = 0$$

Consequently, the probability that a random variable is "greater than" or "greater than or equal to" a number is the same in for continuous random variables. The same is true of" less than" and "less than or equal to" signs for continuous random variables. This equivalence is absolutely not true for discrete random variables.

$$P(a < x < b) = P(a \leq x \leq b) \text{ and } P(a > x > b) = P(a \geq x \geq b)$$

Also it is important to note that substitution of a value into the PDF gives a probability only for a discrete random variable, in order words $P(x = a) = f(a)$ for discrete PDFs only. For a continuous random variable, $f(a)$ by itself doesn't provide a probability. Only the integral of $f(a)$ provides a probability from a continuous random variable.

Example 2.7.: A probability density function has the form

$$f(x) = \begin{cases} \dfrac{x^2}{3} & \text{for } -1 < x < 2 \\ 0 & \text{otherwise} \end{cases}$$

A plot of the probability density distribution is shown in Figure 2.2. This plot is the continuous analog of the discrete histogram.

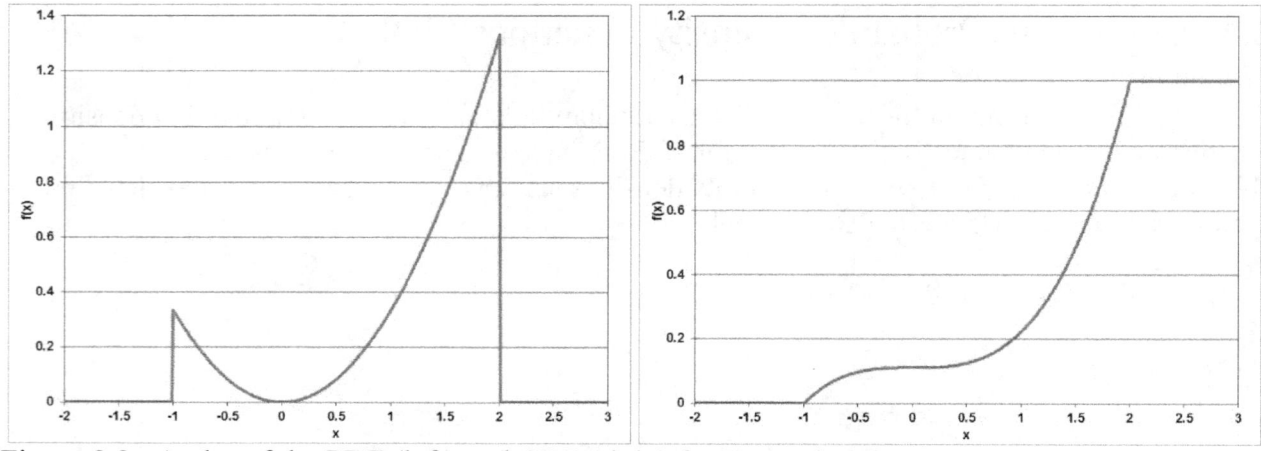

Figure 2.2. A plot of the PDF (left) and CDF (right) for Example 2.7.

The probability of finding an x between a and b is by (equation 2.4)

$$P(a < x < b) = \int_a^b f(x)dx = \begin{cases} \int_a^b \dfrac{x^2}{3} dx & \text{for } -1 < x < 2 \\ \int_a^b 0\, dx & \text{otherwise} \end{cases}$$

$$P(a < x < b) = \int_a^b f(x)dx = \begin{cases} \left(\dfrac{b^3}{9} - \dfrac{a^3}{9}\right) & \text{for } -1 < x < 2 \\ 0 & \text{otherwise} \end{cases}$$

(a) Find $P(-1 < x < 2)$

$$P(-1 < x < 2) = \int_{-1}^{2} f(x)dx = \left(\dfrac{2^3}{9} - \dfrac{(-1)^3}{9}\right) = 1$$

This result makes sense since the PDF is normalized and we have integrated over the entirety of the non-zero range of the random variable.

(b) Find $P(-\infty < x < \infty)$

We cannot integrate over discontinuities in a function. Therefore, we must break-up the integral over continuous parts.

$$P(-\infty < x < \infty) = \int_{-\infty}^{-1} f(x)dx + \int_{-1}^{2} f(x)dx + \int_{2}^{\infty} f(x)dx = 0 + \left(\frac{2^3}{9} - \frac{(-1)^3}{9}\right) + 0 = 1$$

Here we see that actually it is not practically necessary to integrate over the parts of the function where $f(x)=0$, because the integral over those ranges is also 0. In general practice, we just need to perform the integration over those ranges where the PDF, $f(x)$, is non-zero.

(c) Find $P(-\infty < x < 0)$

$$P(-\infty < x < 0) = \int_{-\infty}^{-1} f(x)dx + \int_{-1}^{0} f(x)dx = 0 + \left(\frac{0^3}{9} - \frac{(-1)^3}{9}\right) = \frac{1}{9}$$

Again, it is not necessary to explicitly integrate over anything but the non-zero portions of the PDF, as all other portions contribute nothing to the integral.

(d) Find $P(0 < x < 1)$

$$P(0 < x < 1) = \int_{0}^{1} f(x)dx = \left(\frac{1^3}{9} - \frac{0^3}{9}\right) = \frac{1}{9}$$

Testing a continuous PDF for legitimacy

If you are asked to determine if a given PDF is legitimate, you are required to verify the three criteria in equation (2.4). Generally, the third criterion is given in the problem statement, so you only have to check the first 2 criteria.

The first criteria, $f(x) \geq 0$, can most easily be verified by plotting $f(x)$ and showing that it is never negative. The second criteria, $\int_{-\infty}^{\infty} f(x)dx = 1$, can most easily be verified by direct integration of $f(x)$.

Normalizing a continuous PDF

Continuous PDF's must satisfy $\int_{-\infty}^{\infty} f(x)dx = 1$. Sometimes, you have the functional form of the PDF and you simply need to force it to satisfy this criterion. In that case you need to normalize the PDF so that it sums to unity.

If $f^*(x)$ is an unnormalized PDF, then it can be normalized by the multiplication of a constant,

$$f(x) = cf^*(x) = \frac{1}{\int_{-\infty}^{\infty} f^*(x)dx} f^*(x) \qquad (2.5)$$

where that constant is the inverse of the sum of the unnormalized PDF.

Example 2.8.: Find the value of c that normalizes the PDF.

$$f(x) = \begin{cases} cx^2 & \text{for } -1 < x < 2 \\ 0 & \text{otherwise} \end{cases}$$

To normalize:

$$\int_{-\infty}^{\infty} f(x)dx = 1 = \int_{-1}^{2} cx^2 dx = c\frac{x^3}{3}\Big|_{-1}^{2} = c\left(\frac{9}{3}\right) = 3c$$

$$c = \frac{1}{3}$$

So the normalized PDF is

$$f(x) = \begin{cases} \dfrac{x^2}{3} & \text{for } -1 < x < 2 \\ 0 & \text{otherwise} \end{cases}$$

Continuous Cumulative distributions

The cumulative distribution $F(x)$ of a continuous random variable x with density function $f(x)$ is

$$F(a) = P(x \le a) = \int_{-\infty}^{a} f(x)dx \quad \text{for } -\infty < x < \infty$$

(5.6)

This function gives the probability that a randomly selected value of the variable x is less than a. The implicit lower limit of cumulative distribution is negative infinity.

$$F(a) = P(-\infty \le x \le a) = P(x \le a)$$

Example 2.9.: Determine the cumulative distribution function for the PDF of Example 2.7.

$$F(a) = P(x \le a) = \int_{-\infty}^{a} f(t)dt = \begin{cases} 0 & \text{for } x < -1 \\ \left(\dfrac{a^3}{9} + \dfrac{1^3}{9}\right) & \text{for } -1 < x < 2 \\ 1 & \text{for } x > 2 \end{cases}$$

A plot of the cumulative distribution function for the PDF of Example 2.7. is shown in Figure 2.2. The CDF is again monotonically increasing. It begins at zero and ends at unity, since the PDF is normalized.

2.5. Relations between Inequalities

In the above section we have defined a specific function for the probability that x is less than or equal to a, namely the cumulative distribution. But what about when x is greater than a, or strictly less than a, etc.? Here, we discuss those possibilities.

Consider the fact that the probability of all outcomes must sum to one. Then we can write (regardless of whether the PDF is discrete or continuous)

$$P(x < a) + P(x = a) + P(x > a) = 1$$

Using the union rule we can write:

$$P(x \le a) = P[(x < a) \cup (x = a)] = P(x < a) + P(x = a) + P[(x < a) \cap (x = a)]$$

The intersection is zero, because x cannot equal a and be less than a, so

$$P(x \le a) = P[(x < a) \cup (x = a)] = P(x < a) + P(x = a)$$

Similarly

$$P(x \ge a) = P[(x > a) \cup (x = a)] = P(x > a) + P(x = a)$$

Using these three rules, we can create a generalized method for obtaining any arbitrary probability. On the other hand, we can use the rules to create way to obtain any probability from just the cumulative distribution function. (This will be important later when we use PDF's for which only the cumulative distribution function is given.) Regardless of which method you use, you will obtain the same answer. In Table 2.1, we summarize the expression of each probability in terms of the cumulative PDF.

The continuous case has one important difference. In the continuous case, the probability of a random variable x equaling a single value a is zero. Why? Because the probability is a ratio of the number of ways of getting a over the total number of ways in the sample space. There is only one way to get a, namely $x=a$. But in the denominator, there is an infinite number of values of x, since x is continuous. Therefore, the $P(x=a)=0$. We can show this using the definition is we write,

$$P(x = a) = \int_a^a f(x)dx = 0 \quad \text{for continuous PDF's only.}$$

One consequence of this is that

$$P(x \leq a) = P(x < a) + P(x = a) = P(x < a)$$

$$P(x \geq a) = P(x > a) + P(x = a) = P(x > a)$$

The probability of $x \leq a$ is the same as $x < a$. Likewise, the probability of $x \geq a$ is the same as $x > a$. This fact makes the continuous case easy to generate. In Table 2.2, we summarize the expression of each probability in terms of the cumulative PDF.

Probability	Definition	from cumulative PDF
$P(x = a)$	$f(a)$	$P(x \leq a) - P(x \leq a - 1)$
$P(x \leq a)$	$\sum_{x \leq a} f(x)$	$P(x \leq a)$
$P(x < a)$	$\sum_{x < a} f(x)$	$P(x \leq a - 1)$
$P(x \geq a)$	$\sum_{x \geq a} f(x)$	$1 - P(x \leq a - 1)$
$P(x > a)$	$\sum_{x > a} f(x)$	$1 - P(x \leq a)$

Table 2.1. Relations between inequalities for <u>discrete</u> random variables.

Probability	Definition	from cumulative PDF
$P(x = a)$	$\int_a^a f(x)dx = 0$	0
$P(x \leq a)$ or $P(x < a)$	$\int_{-\infty}^a f(x)dx$	$P(x \leq a)$
$P(x \geq a)$ or $P(x > a)$	$\int_a^\infty f(x)dx$	$1 - P(x \leq a)$

Table 2.2. Relations between inequalities for <u>continuous</u> random variables.

Let's close out this section with two more vocabulary words used to describe PDFs. The definitions of symmetric and skewed distributions are provided below. An example of each are plotted in Figure 2.3.

Symmetric
A probability density distribution is said to be symmetric if it can be folded along a vertical axis so that the two sides coincide.

Skew
A probability density distribution is said to be skewed if it is not symmetric.

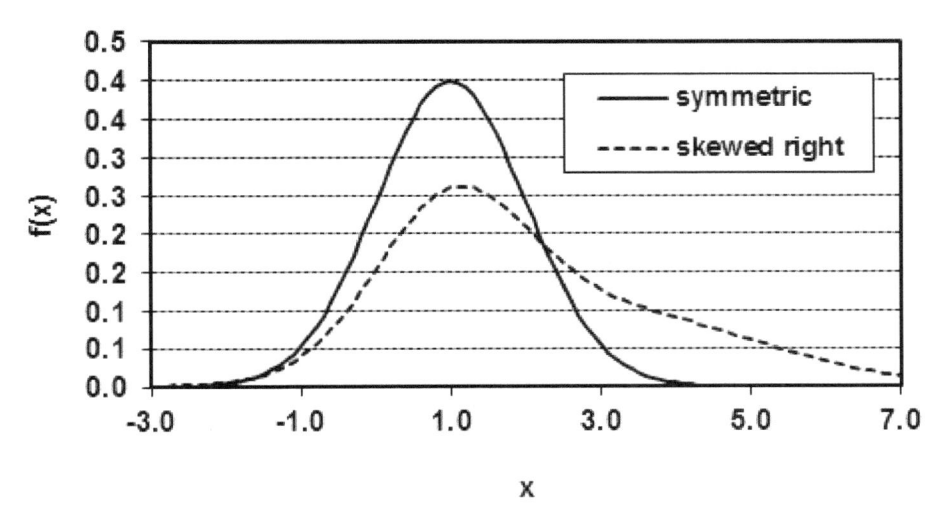

Figure 2.3. Examples of symmetric and skewed PDFs.

2.6. Discrete Joint Probability Distribution Functions

Thus far in this chapter, we have assumed that we have only one random variable. In many practical applications there are more than one random variable. The behavior of sets of random variables is described by Joint PDFs. In this book, we explicitly extend the formalism to two random variables. It can be extended to an arbitrary number of random variables. We will present this extension twice, once for discrete random variables and once for continuous random variables.

The function $f(x,y)$ is a joint probability distribution or probability mass function of the discrete random variable X and Y if

$$f(x,y) \geq 0$$
$$\sum_x \sum_y f(x,y) = 1 \quad (2.7)$$
$$P(x = a \cap y = b) = f(a,b)$$

This is just the two variable extension of equation (2.1). The PDF is always positive. The PDF is normalized, summing to unity, over all combinations of x and y. The Joint PDF gives the intersection of the probability.

The extension of the cumulative discrete probability distribution, equation (2.4), is that for any region A in the x-y plane,

$$F(a,b) = P(x \leq a \cap y \leq b) = \sum_{x \leq a} \sum_{y \leq b} f(x,y) \tag{2.8}$$

That is to say, the probability that a result (x,y) is inside an arbitrary area, A, is equal to the sum of the probabilities for all of the discrete events inside A.

Example 2.10.: Consider the discrete Joint PDF, $f(x,y)$, as given in the table below.

y \ x	1	2	3
0	1/20	2/20	3/20
1	4/20	1/20	2/20
2	2/20	2/20	3/20

Compute the probability that x is 1 and y is 2.

$$P(x=1 \cap y=2) = f(1,2) = \frac{2}{20}$$

Compute the probability that x is less than or equal to 1 and y is less than or equal to 2.

$$P(x \leq 1 \cap y \leq 2) = \sum_{x=1}^{1}\sum_{y=0}^{2} f(x,y) = f(1,0) + f(1,1) + f(1,2) = \frac{1}{20} + \frac{4}{20} + \frac{2}{20} = \frac{7}{20}$$

2.7. Continuous Joint Probability Density Functions

The distribution of continuous variables can be extended in an exactly analogous manner as was done in the discrete case. The function $f(x,y)$ is a Joint Density Function of the continuous random variables, x and y, if

Random Variables and Probability Distributions - 39

$$f(x, y) \geq 0 \quad \text{for all } x, y \in R$$

$$\int_{-\infty}^{\infty} \int_{-\infty}^{\infty} f(x, y) dx dy = 1 \tag{2.9}$$

$$P[(x, y) \in A] = \iint_A f(x, y) dx dy$$

This is just the two variable extension of equation (2.4). The PDF is always positive. The PDF is normalized, integrating to unity, over all combinations of x and y. The Joint PDF gives the intersection of the probability. That third equation takes a specific form, depending on the shape of the Area A. For a rectangle, it would look like:

$$P(a < x < b \cap c < y < d) = \int_c^d \int_a^b f(x, y) dx dy$$

Naturally, the cumulative distribution of the single variable case can also be extended to 2-variables.

$$F(x, y) = P(x \leq a \cap y \leq b) = \int_{-\infty}^b \int_{-\infty}^a f(x, y) dx dy \tag{2.10}$$

Example 2.11.: Given the continuous Joint PDF, find $P(0 \leq x \leq 0.5 \cap 0.5 \leq y \leq 1)$

$$f(x, y) = \begin{cases} \frac{2}{5}(2x + 3y) & \text{for } 0 \leq x \leq 1, 0 \leq y \leq 1 \\ 0 & \text{otherwise} \end{cases}$$

$$P\left(0 < x < \frac{1}{2} \cap \frac{1}{2} < y < 1\right) = \int_{\frac{1}{2}}^{1} \int_0^{\frac{1}{2}} \frac{2}{5}(2x + 3y) dx dy$$

$$= \int_{0.5}^{1} \left[\frac{2x^2}{5} + \frac{6xy}{5}\right]_0^{0.5} dy = \int_{0.5}^{1} \left(\frac{1}{10} + \frac{3y}{5}\right) dy = \left[\frac{y}{10} + \frac{3y^2}{10}\right]_{0.5}^{1} = \frac{11}{40}$$

At this point, we should point out two things. First, we have presented four cases for discrete and continuous PDFs for one or two random variables. There are really only two core equations, the requirements for the probability distribution and the definition of the cumulative probability distribution. We have shown these 2 equations for 4 cases; (i) discrete, one variable, (ii) continuous one variable, (iii) discrete, two variable, and (iv) continuous 2 variable. Re-examine these eight equations to make sure that you see the similarities.

In this text, we are stopping at two variables. However, discrete and continuous probability distributions can be functions of an arbitrary number of variables.

2.8. Marginal Distributions and Conditional Probabilities

Marginal distributions give us the probability of obtaining one variable outcome regardless of the value of the other variable. Marginal distributions are needed to calculate conditional probabilities. The marginal distributions of x alone and of y alone are

$$g(x) = \sum_y f(x,y) \quad \text{and} \quad h(y) = \sum_x f(x,y) \tag{2.11}$$

for the discrete case and

$$g(x) = \int_{-\infty}^{\infty} f(x,y) dy \quad \text{and} \quad h(y) = \int_{-\infty}^{\infty} f(x,y) dx \tag{2.12}$$

for the continuous case.

Example 2.12.: The discrete joint density function is given by the following table.

y \ x	1	2	3
0	1/20	2/20	3/20
1	4/20	1/20	2/20
2	2/20	2/20	3/20

Compute the marginal distribution of x at all possible values of x:

$g(x=0) = f(0,1) + f(0,2) + f(0,3) = 6/20$
$g(x=1) = f(1,1) + f(1,2) + f(1,3) = 7/20$
$g(x=2) = f(2,1) + f(2,2) + f(2,3) = 7/20$

Compute the marginal distribution of y at all possible values of y:

$h(y=1) = f(0,1) + f(1,1) + f(2,1) = 7/20$
$h(y=2) = f(0,2) + f(1,2) + f(2,2) = 5/20$
$h(y=3) = f(0,3) + f(1,3) + f(2,3) = 8/20$

We note that both marginal distributions are legitimate PDFs and satisfy the three requirements of equation (2.1), namely that they are non-negative, normalized and their evaluation yields probabilities.

Random Variables and Probability Distributions - 41

Example 2.13.: The continuous joint density function is

$$f(x,y) = \begin{cases} \frac{2}{5}(2x+3y) & \text{for } 0 \leq x \leq 1, 0 \leq y \leq 1 \\ 0 & \text{otherwise} \end{cases}$$

Find g(x) and h(y) for this joint density function.

$$g(x) = \int_{-\infty}^{\infty} f(x,y)dy = \int_{-\infty}^{0} 0\,dy + \int_{0}^{1} \frac{2}{5}(2x+3y)dy + \int_{1}^{\infty} 0\,dy$$

$$= \frac{2}{5}\left(2xy + \frac{3}{2}y^2\right)\Big|_0^1 = \frac{4x}{5} + \frac{3}{5}$$

$$h(y) = \int_{-\infty}^{\infty} f(x,y)dx = \int_{-\infty}^{0} 0\,dx + \int_{0}^{1} \frac{2}{5}(2x+3y)dx + \int_{1}^{\infty} 0\,dx$$

$$= \frac{2}{5}\left(x^2 + 3yx\right)\Big|_0^1 = \frac{2}{5} + \frac{6y}{5}$$

These marginal distributions themselves satisfy all the properties of a probability density distribution, namely the requirements in equation (2.4). The physical meaning of the marginal distribution functions are that they give the individual effects of *x* and *y* separately.

Conditional Probability

We now relate the conditional probability to the marginal distributions defined above. We do this first for the discrete case and then for the continuous case.

Let *x* and *y* be two discrete random variables. The conditional distribution of the random variable *y=b*, given that *x=a*, is

$$f(y=b|x=a) = \frac{f(x=a, y=b)}{g(x=a)} \quad \text{where } g(a) > 0 \tag{2.13}$$

Similarly, the conditional distribution of the random variable *x=a*, given that *y=b*, is

$$f(x=a|y=b) = \frac{f(x=a, y=b)}{h(y=b)} \quad \text{where } h(b) > 0 \tag{2.14}$$

You should see that this conditional distribution is simply the application of the definition of the conditional probability, which we learned in Chapter 1,

$$P(B\mid A) = \frac{P(A\cap B)}{P(A)} \quad \text{for } P(A) > 0 \tag{1.38}$$

Example 2.14.: Given the discrete PDF in Example 2.12., calculate
(a) $f(y = 2\mid x = 2)$
(b) $f(x = 1\mid y \le 2)$

(a) $f(y = 2\mid x = 2)$. Using the conditional probability definition:

$$f(y = b\mid x = a) = \frac{f(x = a, y = b)}{g(x = a)}$$

We already have the denominator: $g(x=2) = 7/20$. The numerator is $f(x=2, y=2) = 2/20$. Therefore, the conditional probability is:

$$f(y = 2\mid x = 2) = \frac{2/20}{7/20} = \frac{2}{7}$$

(b) $f(x = 1\mid y \le 2)$

$$f(x = 1\mid y \le 2) = \frac{f(x = 1, y \le 2)}{h(y \le 2)}$$

The numerator is the sum over all values of $f(x,y)$ for which $x=1$, and $y \le 2$. So

$$f(x = 1, y \le 2) = f(1,0) + f(1,1) + f(1,2) = \frac{1}{20} + \frac{4}{20} + \frac{2}{20} = \frac{7}{20}$$

The denominator is the sum over all $h(y)$ for $y \le 2$

$$h(y \le 2) = h(1) + h(2) = \frac{7}{20} + \frac{5}{20} = \frac{12}{20}$$

Therefore,

$$f(x = 1\mid y \le 2) = \frac{7/20}{12/20} = \frac{7}{12}$$

Random Variables and Probability Distributions - 43

A similar treatment can be done for the continuous case. Let x and y be two continuous variables. The conditional distribution of the random variable $c<y<d$, given that $a<x<b$, is

$$P(c < y < d | a < x < b) = \frac{\int_c^d \int_a^b f(x,y)dxdy}{\int_a^b g(x)dx} \quad \text{where } \int_a^b g(x)dx > 0 \qquad (2.15)$$

Similarly, the conditional distribution of the random variable $a<x<b$, given that $c<y<d$, is

$$P(a < x < b | c < y < d) = \frac{\int_c^d \int_a^b f(x,y)dxdy}{\int_c^d h(y)dy} \quad \text{where } \int_c^d h(y)dy > 0 \qquad (2.16)$$

Example 2.15.: Consider the continuous joint PDF in problem 2.13. Calculate $P(0 < X < 0.5 | 0.5 < y < 1)$.

$$P(0 < X < 0.5 | 0.5 < y < 1) = \frac{P(0 < X < 0.5 \cap 0.5 < y < 1)}{P(0 < X < 0.5)}$$

$$P(0 < X < 0.5 | 0.5 < y < 1) = \frac{\int_{0.5}^{1} \int_0^{0.5} f(x,y)dxdy}{\int_{0.5}^{1} h(y)dy}$$

We calculated the numerator in Example 2.11. and it had a numerical value of 11/40.

The denominator is:

$$\int_{0.5}^{1} h(y)dy = \int_{0.5}^{1} \left(\frac{2}{5} + \frac{6y}{5}\right)dy = \left[\frac{2y}{5} + \frac{6y^2}{10}\right]_{0.5}^{1} = \frac{13}{20}$$

The conditional probability is then

$$P(0 < X < 0.5 | 0.5 < y < 1) == \frac{\frac{11}{40}}{\frac{13}{20}} = \frac{11}{26}$$

2.9. Statistical Independence

In Chapter 1, we used the conditional probability rule to as a check for independence of two outcomes. This same approach is repeated here for two random variables. Let x and y be two random variables, discrete or continuous, with joint probability distribution $f(x,y)$ and marginal distributions $g(x)$ and $h(y)$. The random variables x and y are said to be statistically independent iff (if and only if)

$$f(x,y) = g(x)h(y) \quad \text{if and only if } x \text{ and } y \text{ are independent} \tag{2.17}$$

for all possible values of (x,y). This should be compared with the rule for independence of probabilities:

$$P(A \cap B) = P(A)P(B) \quad \text{iff and } A \text{ and } B \text{ are independent events} \tag{1.44}$$

Example 2.16.: In the continuous example given above, determine whether x and y are statistically independent random variables.

$$f(x,y) = \begin{cases} \frac{2}{5}(2x+3y) & \text{for } 0 \leq x \leq 1, 0 \leq y \leq 1 \\ 0 & \text{otherwise} \end{cases}$$

$$g(x) = \frac{4x}{5} + \frac{3}{5} \quad \text{and} \quad h(y) = \frac{2}{5} + \frac{6y}{5}$$

$$g(x)h(y) = \left(\frac{4x}{5} + \frac{3}{5}\right)\left(\frac{2}{5} + \frac{6y}{5}\right) = \frac{1}{25}(8x + 24xy + 6 + 18y)$$

The product of marginal distributions is not equal to the joint probability density distribution. Therefore, the variables are not statistically independent.

2.10. Problems

Problem 2.1.
Determine the value of c so that the following functions can serve as a PDF of the discrete random variable X.

$$f(x) = c(x^2 + 4) \text{ where } x = 0,1,2,3;$$

Problem 2.2.
A shipment of 7 computer monitors contains 2 defective monitors. A business makes a random purchase of 3 monitors. If x is the number of defective monitors purchased by the company, find the probability distribution of X. (This means you need three numbers, f(x=0), f(x=1), and f(x=2) because the random variable, X = number of defective monitors purchased, has a range from 0 to 2. Also, find the cumulative PDF, F(x). Plot the PDF and the cumulative PDF. These two plots must be turned into class on the day the homework is due.

Problem 2.3.
A continuous random variable, X, that can assume values between x=2 and x=5 has a PDF given by

$$f(x) = \frac{2}{27}(1+x)$$

Find (a) P(X<4) and find (b) P(3<X<4). Plot the PDF and the cumulative PDF.

Problem 2.4.
Consider a system of particles that sit in an electric field where the energy of interaction with the electric field is given by E(x) = 2477.572 + 4955.144x, where x is spatial position of the particles. The probability distribution of the particles is given by statistical mechanics to be f(x) = c*exp(-E(x)/(R*T)) for 0<x<1 and 0 otherwise, where R = 8.314 J/mol/K and T = 270.0 Kelvin.
  (a) Find the value of c that makes this a legitimate PDF.
  (b) Find the probability that a particles sits at x<0.25
  (c) Find the probability that a particles sits at x>0.75
  (d) Find the probability that a particles sits at 0.25<x<0.75

Problem 2.5.
Let X denote the reaction time, in seconds, to a certain stimulant and Y denote the temperature (reduced units) at which a certain reaction starts to take place. Suppose that the random variables X and Y have the joint PDF,

$$f(x,y) = \begin{cases} cxy & \text{for } 0 < x < 1; 0 < y < 2.1 \\ 0 & \text{elsewhere} \end{cases}$$

where c = 0.907029. Find (a) $P\left(0 \le X \le \frac{1}{2} \text{ and } \frac{1}{4} \le Y \le \frac{1}{2}\right)$ and (b) $P(X < Y)$.

Problem 2.6.
Let X denote the number of times that a control machine malfunctions per day (choices: 1, 2, 3) and Y denote the number of times a technician is called. f(x,y) is given in tabular form.

f(x,y)		x	1	2	3
		1	0.05	0.05	0.1
y		2	0.05	0.1	0.35
		3	0.0	0.2	0.1

(a) Evaluate the marginal distribution of X.
(b) Evaluate the marginal distribution of Y.
(c) Find P(Y = 3|X = 2).

Chapter 3. Expectations

3.1. Introduction

In this chapter, we define five mathematical expectations—the mean, variance, standard deviation, covariance and correlation coefficient. We apply these general formula to an array of situations involving discrete and continuous random variables obeying single and joint probability distribution functions in order to evaluate expectations of both random variables and functions of random variables. Ideally, the reader observes the common analogy in the application of the five concepts expressed in a variety of different ways.

3.2. Mean of a Random Variable

The "mean" is another name for the "average". A third synonym for mean is the "expected value". Let x be a random variable with probability distribution $f(x)$. The mean of x is

$$\mu_x = E(x) = \sum_x x f(x) \qquad (3.1.a)$$

if x is a discrete random variable, and

$$\mu_x = E(x) = \int_{-\infty}^{\infty} x f(x) dx \qquad (3.1.b)$$

if x is a continuous random variable. We observe that this expectation weights each value of x by its corresponding probability.

Example 3.1: You all may be pretty upset that I suggest that the complicated formulae above give the average when you have been taught since elementary school that the average is given by:

$$\mu_x = \frac{1}{n} \sum_{i=1}^{n} x_i$$

where you just sum up all the elements in the set and divide by the number of elements. Well, let me reassure you that what you learned in elementary school is not wrong. This formula above is one case of equation (3.1) where the probability distribution, $f(x) = 1/n$. (This is called the discrete uniform probability distribution in the following chapter.) Since, in this case, $f(x)$ is not a function of x, it can be pulled out of the summation, giving the familiar result for the mean. However, the uniform distribution is just one of an infinite number of probability distributions. The general formula will apply for any probability distribution.

3.3. Mean of a Function of a Random Variable

Equation 3.1 gives the expected value of the random variable. In general, however, we need the expected value of a function of that random variable. In the general case, our equations which define the mean become:

$$\mu_{h(x)} = E(h(x)) = \sum_x h(x)f(x) \qquad (3.2.a)$$

if x is discrete, and

$$\mu_{h(x)} = E(h(x)) = \int_{-\infty}^{\infty} h(x)f(x)dx \qquad (3.2.b)$$

if x is continuous, where $h(x)$ is some arbitrary function of x. You should see that equation (3.1) is an example of the case of equation (3.2) where $h(x)=x$. This is one kind of function. However, there is no point in learning a formula for one function, when the formula for all functions is at hand. So, equation (3.2) is the equation to remember.

Example 3.2: In a gambling game, three coins are tossed. A man is paid $5 when all three coins turn up the same, and he will lose $3 otherwise. What is the expected gain?

In this problem, the random variable, x, is the number of heads. The distribution function, $f(x)$ is a uniform distribution for the 8 possible outcomes of the gambling game. The function $h(x)$ is the payout or forfeit for outcome x.

outcome	x	h(x)	f(x)
HHH	3	+$5	1/8
HHT	2	-$3	1/8
HTH	2	-$3	1/8
HTT	1	-$3	1/8
THH	2	-$3	1/8
THT	1	-$3	1/8
TTH	1	-$3	1/8
TTT	0	+$5	1/8

We know that the probability distribution $f(x) = 1/8$, since there are 8 random, equally probable outcomes. Using equation (3.2), we find:

$$\mu_{h(x)} = \sum_x h(x)f(x) = \sum_{i=1}^{8} h(x)\left(\frac{1}{8}\right) = \frac{1}{8}\sum_{i=1}^{8} h(x) = \frac{1}{8}(5-3-3-3-3-3-3+5) = -1$$

The average outcome is that the gambler loses a dollar.

We can also work this same problem another way and make the distribution over the number of heads rather than all possible outcomes. In this case, the table looks like:

outcome	x	h(x)	f(x)
HHH	3	+$5	1/8
HHT,HTH,THH	2	-$3	3/8
TTH, THT, TTH	1	-$3	3/8
TTT	0	+$5	1/8

Then using equation (3.2) again, we find:

$$\mu_{h(x)} = \sum_x h(x)f(x) = \sum_{i=1}^{4} h(x)f(x) = \left[(5)\frac{1}{8}+(-3)\frac{3}{8}+(-3)\frac{3}{8}+(5)\frac{1}{8}\right] = -1$$

In fact there are other ways to define the problem. Choose a distribution that makes sense to you. They will all give the same answer so long as your distribution agrees with the physical reality of the problem.

Example 3.3: Consider the continuous random variable describing the temperature in a furnace, T, which is distributed between 500 K and 1000 K according to a PDF is given by $f(T) = 1000/T^2$. The diffusivity is given as a function of temperature, $D(T) = 5\exp(-800/T)$ cm²/s. What is the average value of temperature? What is the average value of the diffusivity? Can you obtain the average value of the diffusivity by substituting in the average value of the temperature in the formula for the diffusivity?

The average temperature (the random variable) is computed via equation (3.1.b)

$$\mu_T = \int_{500}^{1000} Tf(T)dt = \int_{500}^{1000} T\frac{1000}{T^2}dt = 1000\int_{500}^{1000}\frac{1}{T}dt = 1000\ln(T)\Big|_{500}^{1000} = 1000\ln(2) \approx 693.147\,\text{K}$$

The average diffusivity (a function of the random variable) is computed via equation (3.2.b)

$$\mu_{D(T)} = \int_{500}^{1000} D(T)f(T)dT = \int_{500}^{1000} 5\exp(-800/T)\frac{1000}{T^2}dT = 5000\int_{500}^{1000}\exp(-800/T)\frac{1}{T^2}dT$$

$$= \frac{5000}{800}\left[\exp(-1200/T)\right]_{500}^{1000} \approx 1.546\ \frac{cm^2}{s}$$

The diffusivity evaluated at the average temperature is

$$D(\overline{T}) = 5\exp(-800/\overline{T}) = 5\exp(-1200/693.147) \approx 1.577\ cm^2/s$$

Therefore, the mean of the diffusivity cannot be computed by simply evaluating the diffusivity at the mean of the temperature. This is true because the diffusivity is a nonlinear function of the temperature.

3.4. Mean of a Function of Two Random Variables

Equation (3.2) gives the expected value of a function of one random variable. This equation can be simply extended to a function of two random variables.

$$\mu_{h(x,y)} = E(h(x,y)) = \sum_x \sum_y h(x,y)f(x,y) \qquad (3.3.a)$$

if x and y are discrete random variables, and

$$\mu_{h(x,y)} = E(h(x,y)) = \int_{-\infty}^{\infty}\int_{-\infty}^{\infty} h(x,y)f(x,y)dxdy \qquad (3.3.b)$$

if x and y are continuous random variables. You should see that equation (3.3) is entirely analogous to equation (3.2).

Example 3.4: Given the joint probability density function

$$f(x,y) = \begin{cases} \frac{2}{5}(2x+3y) & \text{for } 0 \le x \le 1, 0 \le y \le 1 \\ 0 & \text{otherwise} \end{cases}$$

find the mean of $h(x,y) = y$.

Using equation (3.3.b) we have

$$\mu_{h(x,y)} = E(h(X,Y)) = \int_{-\infty}^{\infty}\int_{-\infty}^{\infty} h(x,y)f(x,y)dxdy = \int_0^1\int_0^1 y\left(\frac{2(2x+3y)}{5}\right)dxdy$$

$$\mu_{h(x,y)} = \int_0^1\int_0^1 y\left(\frac{2(2x+3y)}{5}\right)dxdy = \int_0^1 \frac{2}{5}\left(yx^2+3y^2x\right)\bigg|_{x=0}^{x=1} dy = \int_0^1 \frac{2}{5}\left(y+3y^2\right)dy$$

$$\mu_{h(x,y)} = \frac{2}{5}\left(\frac{y^2}{2}+y^3\right)\bigg|_{y=0}^{y=1} = \frac{2}{5}\left(\frac{1}{2}+1\right) = \frac{6}{10}$$

What this says is that the average value of *y* for this joint probability distribution function is 0.6.

3.5. Variance of a Random Variable

The mean is one parameter of a distribution of data. It gives us some indication of the location of the random variable. It does not however give us any information about the distribution of the random variable. For example, in Figure 3.1., we observe the visually three PDFs with different values of the mean but the same value of the variance. The location of each PDF is different but the spread of the PDF remains the same. In Figure 3.2., we observe three PDFs with different values of the variance but the same value of the mean. Here, the location of the PDF remains constant but the spread of the PDF increases with increasing variance. The **variance** is a statistical measure of this spread.

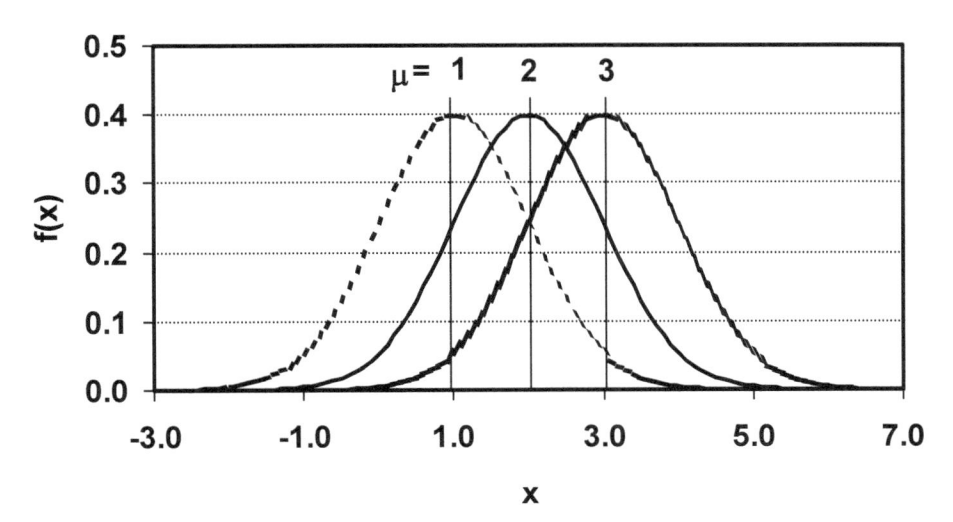

Figure 3.1. Three continuous probability density functions with common variance but different values of the mean.

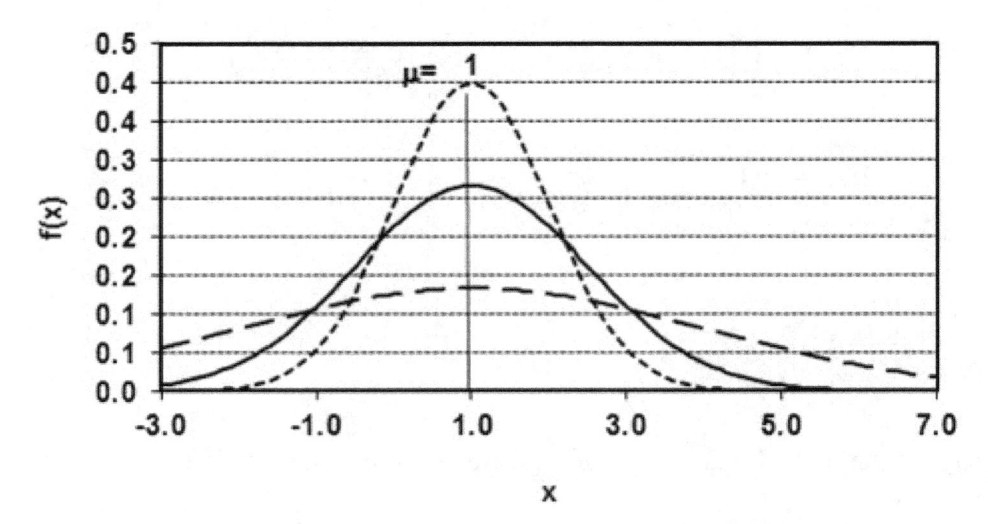

Figure 3.2. Three continuous probability density functions with common mean but different values of the variance.

We define the variance as follows. Let x be a random variable with PDF $f(x)$ and mean μ_x. The **variance** of x, σ_x^2, is defined as

$$\sigma_x^2 = E\left[(x - \mu_x)^2\right] \equiv \sum_x (x - \mu_x)^2 f(x) \qquad (3.4.a)$$

if x is a discrete random variable, and

$$\sigma_x^2 = E\left[(x - \mu_x)^2\right] = \int_{-\infty}^{\infty} (x - \mu_x)^2 f(x) dx \qquad (3.4.b)$$

if x is a continuous random variable. You should see that equation (3.4) is just another case of equation (3.2) where $h(x) = (x - \mu_x)^2$. What the variance gives is "the average of the square of the deviation from the mean". The square is in there so that all the deviations are positive and the variance is a positive number.

Some tricks with the variance:

The definition for the variance given above, if evaluated properly, will always give the correct value of the variance. However, there is another shortcut formula that is often used. We derive the shortcut here.

Expectations - 53

$$\sigma_x^2 = E\left[(x-\mu_x)^2\right] = \int_{-\infty}^{\infty}(x-\mu_x)^2 f(x)dx$$

$$= \int_{-\infty}^{\infty} x^2 f(x)dx - \int_{-\infty}^{\infty} 2x\mu_x f(x)dx + \int_{-\infty}^{\infty} \mu_x^2 f(x)dx$$

$$= \int_{-\infty}^{\infty} x^2 f(x)dx - 2\mu_x \int_{-\infty}^{\infty} xf(x)dx + \mu_x^2 \int_{-\infty}^{\infty} f(x)dx$$

$$= E[x^2] - 2\mu_x E[x] + \mu_x^2$$
$$= E[x^2] - 2\mu_x^2 + \mu_x^2$$
$$= E[x^2] - \mu_x^2 = E[x^2] - E[x]^2$$

Thus we have,

$$\sigma_x^2 = E[x^2] - E[x]^2 \qquad (3.5)$$

People frequently express the variance as "the difference between the mean of the squares and the square of the mean" of the random variable x. They do this because sometimes, you have $E[x^2]$ and $E[x]$ so the variance is often easier to calculate from equation (3.5) than it is from equation (3.4).

Example 3.5.: Calculate the mean and variance of the discrete data set of 10 numbers containing $\{1,2,3,4,5,6,7,8,9,10\}$.

The mean is calculated from equation (3.3.a), where the probability distribution is uniform, i.e., $f(x) = 1/n$. So $\mu_x = 5.5$. The variance is calculated by squaring each number in the set so that you have a new set of x^2 containing $\{1,4,9,16,25,36,49,64,81,100\}$. Then the mean of this set of numbers (using equation (3.3.a)) is $\mu_{x^2} = 38.5$. Now using equation (3.5), we have:

$$\sigma_x^2 = E[x^2] - E[x]^2 = 38.5 - (5.5)^2 = 8.25$$

The variance is always positive. If you don't get a positive answer using this formula then you have most certainly done something wrong.

Warning on using equation (3.5), the shortcut for the variance

The equation 3.5 may look very friendly but it comes with dangers. You often use this equation to obtain a small variance from the difference of two large numbers. Therefore, the answer you obtain may contain round-off errors. You need to keep all your insignificant figures in the averages in order to obtain the variance to the same number of significant figures.

Example 3.6.: Use $\sigma_x^2 = E[x^2] - E[x]^2$ to obtain the variance of the following 10 numbers, using $f(x) = 1/10$.

```
s = [9.92740197834152
  10.06375116530286
   9.98603320980938
  10.07806434475388
  10.04698164235319
  10.03746471832082
   9.96922239354823
   9.93320694775544
   9.93112251526341
   9.93822326228399]
```

Below we give a table that reports the means and variance when keeping different number of significant figures:

4 sig figs: μ_x = 9.990000000e+000 μ_{x^2} = 9.983000000e+001 σ_x^2 = 2.990000000e-002

5 sig figs: μ_x = 9.991000000e+000 μ_{x^2} = 9.982600000e+001 σ_x^2 = 5.919000000e-003

6 sig figs: μ_x = 9.991100000e+000 μ_{x^2} = 9.982630000e+001 σ_x^2 = 4.220790000e-003

7 sig figs: μ_x = 9.991150000e+000 μ_{x^2} = 9.982626000e+001 σ_x^2 = 3.181677500e-003

8 sig figs: μ_x = 9.991147000e+000 μ_{x^2} = 9.982626500e+001 σ_x^2 = 3.246624391e-003

9 sig figs: μ_x = 9.991147200e+000 μ_{x^2} = 9.982626470e+001 σ_x^2 = 3.242327932e-003

10 sig figs: μ_x = 9.991147220e+000 μ_{x^2} = 9.982626473e+001 σ_x^2 = 3.241958286e-003

all sig figs: μ_x = 9.991147218e+000 μ_{x^2} = 9.982626473e+001 σ_x^2 = 3.242000540e-003

You can see that when we only keep 4 significant figures, our calculated variance is off by 822%! You need to keep additional significant figures in the mean and the mean of the squares in order to get the variance with any accuracy.

For your information, the MATLAB code that I used to generate the data is provided in the appendix of this chapter.

3.6. Standard Deviation

The standard deviation, σ_x, is the positive square root of the variance, σ_x^2.

Expectations - 55

Example 3.7.: Calculate the standard deviation of the discrete data set of 10 numbers containing {1,2,3,4,5,6,7,8,9,10}. We calculated the variance in the example 3.5. above. The standard deviation is the square root of the variance

$$\sigma_x = \sqrt{\sigma_x^2} = \sqrt{8.25} = 2.872281323$$

The standard deviation gives us a number *in the same units as the random variable x*, which describes the spread of the data.

3.7. Variance of a Function of a Random Variable

Let x be a random variable with PDF $f(x)$. Let $g(x)$ be an arbitrary function of x. We know that the mean of $g(x)$ is, and mean $\mu_{g(x)}$ from equation (3.2). The **variance** of a function $g(x)$, is

$$\sigma_{g(x)}^2 = E\left[(g(x) - \mu_{g(x)})^2\right] = \sum_x (g(x) - \mu_{g(x)})^2 f(x) \qquad (3.6.a)$$

if x is a discrete random variable and

$$\sigma_{g(x)}^2 = E\left[(g(x) - \mu_{g(x)})^2\right] = \int_{-\infty}^{\infty} (g(x) - \mu_{g(x)})^2 f(x)dx \qquad (3.6.b)$$

if x is a continuous random variable. You should see that equation (3.6) is just another case of equation (3.2) where the function of the random variable is $h(x) = (g(x) - \mu_{g(x)})^2$. As in the case where the function was $h(x) = (x - \mu_x)^2$, (in equation (3.4)), equation (3.6) can also be reduced to a second form:

$$\sigma_{g(x)}^2 = E\left[g(x)^2\right] - E[g(x)]^2 \qquad (3.7)$$

Beware: we have defined 3 functions, $f(x)$, $g(x)$, and $h(x)$. $f(x)$ is the probability distribution. $g(x)$ is the arbitrary function of the random variable that we would like to know about. $h(x)$ is the function with a mean that provides the variance of its argument. In other words, if $h(x) = (g(x) - \mu_{g(x)})^2$ then $\sigma_{g(x)}^2 = \mu_{h(x)}$.

Example 3.8.: Given the probability density function of the continuous random variable, x,

$$f(x) = \begin{cases} \dfrac{x^2}{3} & \text{for } -1 < x < 2 \\ 0 & \text{otherwise} \end{cases}$$

find the variance of $g(x) = x^{-1}$.

We will use equation (3.7). To do so we must find $E\left[g(x)^2\right]$ and $E[g(x)]$. $E[g(x)]$ is the mean of $g(x)$ and can be calculated from the formula for the mean, equation (3.2).

$$\mu_{g(x)} = E(g(x)) = \int_{-\infty}^{\infty} g(x) f(x) dx \qquad (3.2.b)$$

$$E(x^{-1}) = \int_{-1}^{2} x^{-1}\left(\frac{x^2}{3}\right) dx = \left.\frac{x^2}{6}\right|_{x=-1}^{x=2} = \frac{2^2}{6} - \frac{(-1)^2}{6} = \frac{1}{2}$$

Now we repeat the calculation for the square of $g(x)$

$$E\left[\left(x^{-1}\right)^2\right] = E\left[x^{-2}\right] = \int_{-1}^{2} x^{-2}\left(\frac{x^2}{3}\right) dx = \left.\frac{x}{3}\right|_{x=-1}^{x=2} = \frac{2}{3} - \frac{(-1)}{3} = 1$$

Then we substitute into equation (3.7)

$$\sigma^2_{g(x)} = E\left[g(x)^2\right] - E[g(x)]^2 = 1 - \left(\frac{1}{2}\right)^2 = \frac{3}{4}$$

3.8. Variance & Covariance of a Function of Two Random Variables

By analogous methods, we can extend the variance definition to a function of two variables.

$$\sigma^2_{g(x,y)} = E\left[\left(g(x,y) - \mu_{g(x,y)}\right)^2\right] = \sum_x \sum_y \left(g(x,y) - \mu_{g(x,y)}\right)^2 f(x,y) \qquad (3.8.\text{a})$$

if x and y are discrete random variables and

$$\sigma^2_{g(x,y)} = E\left[\left(g(x,y) - \mu_{g(x,y)}\right)^2\right] = \int_{-\infty}^{\infty}\int_{-\infty}^{\infty} (g(x,y) - \mu_{g(x,y)})^2 f(x,y) dx dy \qquad (3.8.\text{b})$$

if x and y are continuous random variables. You should see that equation (3.8) is just another case of equation (3.3) where the function of the random variable, let's call it $h(x,y) = (g(x,y) - \mu_{g(x,y)})^2$. Again, equation (3.8) can be rewritten

$$\sigma^2_{g(x,y)} = E[g(x,y)^2] - E[g(x,y)]^2 = \mu_{h(x,y)} \tag{3.9}$$

Now, let's think about equation 3.8. If $g(x,y)=x$, then $h(x,y)= (x- \mu_x)^2$ and we calculate the variance of x from equation (3.8). If $g(x,y)=y$, then $h(x,y)= (y- \mu_y)^2$ and we have the variance of y from equation (3.8).

Now, if $h(x,y) = (x - \mu_x)(y - \mu_y)$, then we can use equation (3.8) to calculate the **COVARIANCE**, σ_{xy}. The covariance has the units of xy. There is no function $g(x)$ defined for the covariance, so equation (3.9) does not apply to the calculation of the covariance. But if you substitute $h(x,y) = (x - \mu_x)(y - \mu_y)$ into equation (3.2) and solve as we did to arrive with equation (3.5) you find:

$$\sigma_{XY} = E[xy] - E[x]E[y] = \mu_{XY} - \mu_X \mu_Y \tag{3.10}$$

The qualitative significance of the covariance is the dependency between variables x and y.

σ_{XY}	qualitative significance
$\sigma_{XY} > 0$	as x increases, y increases
$\sigma_{XY} = 0$	x and y are independent
$\sigma_{XY} < 0$	as x increases, y decreases

Table 3.1. Qualitative significance of the covariance.

Example 3.9.: Given the joint probability density function

$$f(x,y) = \begin{cases} \frac{2}{5}(2x+3y) & \text{for } 0 \le x \le 1, 0 \le y \le 1 \\ 0 & \text{otherwise} \end{cases}$$

find the covariance of x and y.

To find the covariance, we need: $E[xy]$, $E[x]$ and $E[y]$. We already calculated $E[y]$ in example 3.3. and we found $E[y]=0.6$. Using a similar procedure, we calculate, the expected value of x, $E[x]$

$$E[x] = \int_{-\infty}^{\infty}\int_{-\infty}^{\infty} xf(x,y)dxdy = \int_0^1\int_0^1 x\left(\frac{2(2x+3y)}{5}\right)dxdy$$

$$E[x] = \int_0^1 \frac{2}{5}\left(\frac{2x^3}{3} + 3y\frac{x^2}{2}\right)\bigg|_{x=0}^{x=1} dy = \int_0^1 \frac{2}{5}\left(\frac{2}{3} + \frac{3y}{2}\right)dy$$

$$E[x] = \frac{2}{5}\left(\frac{2y}{3} + \frac{3y^2}{4}\right)\bigg|_{y=0}^{y=1} = \frac{2}{5}\left(\frac{2}{3} + \frac{3}{4}\right) = \frac{17}{30}$$

In an analogous fashion, we calculate $E[xy]$

$$E[xy] = \int_{-\infty}^{\infty}\int_{-\infty}^{\infty} xyf(x,y)dxdy = \int_0^1\int_0^1 xy\left(\frac{2(2x+3y)}{5}\right)dxdy$$

$$E[xy] = \int_0^1 \frac{2y}{5}\left(\frac{2x^3}{3} + 3y\frac{x^2}{2}\right)\bigg|_{x=0}^{x=1} dy = \int_0^1 \frac{2y}{5}\left(\frac{2}{3} + \frac{3y}{2}\right)dy$$

$$E[xy] = \frac{2}{5}\left(\frac{2y^2}{6} + \frac{3y^3}{6}\right)\bigg|_{y=0}^{y=1} = \frac{2}{5}\left(\frac{2}{6} + \frac{3}{6}\right) = \frac{1}{3}$$

so using equation (3.10), we find:

$$\sigma_{XY} = E[xy] - E[x]E[y] = \frac{1}{3} - \frac{17}{30}\frac{6}{10} = \frac{100-102}{300} = -\frac{1}{150} = -0.006667$$

3.9. Correlation Coefficients

The magnitude of σ_{XY} does not say anything regarding the strength of the relationship between x and y because σ_{XY}. depends on the values taken by x and y. A scaled version of the covariance, called the correlation coefficient is much more useful. The correlation coefficient is defined as

$$\rho_{XY} = \frac{\sigma_{XY}}{\sigma_X \sigma_Y} \tag{3.11}$$

Expectations - 59

This variable ranges from -1 to 1 and is 0 when σ_{XY} is zero. A negative correlation coefficient means that when *y* increases, *x* decreases and vice versa. A positive correlation coefficient means that when *x* increases, *y* also increases, and vice versa for decreasing.

ρ_{XY}	qualitative significance
$\rho_{XY} = 1$	$x = y$
$\rho_{XY} > 0$	as *x* increases, *y* increases
$\rho_{XY} = 0$	*x* and *y* are independent
$\rho_{XY} < 0$	as *x* increases, *y* decreases
$\rho_{XY} = -1$	$x = -y$

Table 3.2. Qualitative significance of the correlation coefficient.

Example 3.11.: Given the joint PDF in example 3.9., find the correlation coefficient and make a statement about whether *x* is strongly or weakly correlated to *y*, relative to the variance of *x* and *y*.

To do this, we need the variance of *x* and *y*, which means we need $E[x^2]$ and $E[y^2]$

$$E[x^2] = \int_{-\infty}^{\infty}\int_{-\infty}^{\infty} x^2 f(x,y)\,dxdy = \int_0^1\int_0^1 x^2\left(\frac{2(2x+3y)}{5}\right)dxdy$$

$$E[x^2] = \int_0^1 \frac{2}{5}\left(\frac{x^4}{2} + yx^3\right)\bigg|_{x=0}^{x=1} dy = \int_0^1 \frac{2}{5}\left(\frac{1}{2} + y\right)dy$$

$$E[x^2] = \frac{2}{5}\left(\frac{y}{2} + \frac{y^2}{2}\right)\bigg|_{y=0}^{y=1} = \frac{2}{5}\left(\frac{1}{2} + \frac{1}{2}\right) = \frac{2}{5}$$

so

$$\sigma_X^2 = E[x^2] - E[x]^2 = \frac{2}{5} - \left(\frac{17}{30}\right)^2 = 0.0789$$

Next we calculate the variance of *y*.

$$E[y^2] = \int_{-\infty}^{\infty}\int_{-\infty}^{\infty} g(x,y)f(x,y)\,dxdy = \int_0^1\int_0^1 y^2\left(\frac{2(2x+3y)}{5}\right)dxdy$$

$$E[y^2] = \int_0^1\int_0^1 y^2\left(\frac{2(2x+3y)}{5}\right)dxdy = \int_0^1 \frac{2}{5}(y^2x^2+3y^3x)\Big|_{x=0}^{x=1} dy = \int_0^1 \frac{2}{5}(y^2+3y^3)dy$$

$$E[y^2] = \frac{2}{5}\left(\frac{y^3}{3}+\frac{3y^4}{4}\right)\Big|_{y=0}^{y=1} = \frac{2}{5}\left(\frac{1}{3}+\frac{3}{4}\right) = \frac{13}{30}$$

so

$$\sigma_Y^2 = E[y^2] - E[y]^2 = \frac{13}{30} - \left(\frac{6}{10}\right)^2 = 0.0733$$

so the correlation coefficient is

$$\rho_{XY} = \frac{\sigma_{XY}}{\sigma_X\sigma_Y} = \frac{-0.00667}{\sqrt{0.0789}\sqrt{0.0733}} = -0.0877$$

The small value of the correlation coefficient indicates that the random variables *x* and *y* are not strongly correlated, but they are weakly negatively correlated.

3.10. Means and Variances of linear combinations of Random Variables

These are several rules for means and variances. These rules have their basis in the theory of linear operators. A linear operator L[x] performs some operation on x, such that:

$$L[ax+by] = aL[x]+bL[y] \tag{3.12}$$

where *x* and *y* are variables and *a* and *b* are constants. This is the fundamental rule which all linear operators must follow.

Consider the differential operator: $L[x] = \frac{d}{dt}[x]$. Is it a linear operator? To prove or disprove the linearity of the differential operator, you must substitute it into equation (3.12) to verify it.

$$\frac{d}{dt}[ax+by] \stackrel{?}{=} a\frac{d}{dt}[x]+b\frac{d}{dt}[y]$$

$$\frac{d}{dt}[ax]+\frac{d}{dt}[by] \stackrel{?}{=} a\frac{d}{dt}[x]+b\frac{d}{dt}[y]$$

$$a\frac{d}{dt}[x]+b\frac{d}{dt}[y] = a\frac{d}{dt}[x]+b\frac{d}{dt}[y] \quad \text{This is an identity.}$$

Expectations - 61

So, we have shown that the differential operator is a linear operator. What about the integral operator, $L[x] = \int x \, dt$?

$$\int [ax + by] \, dt \stackrel{?}{=} a \int x \, dt + b \int y \, dt$$

$$\int ax \, dt + \int by \, dt \stackrel{?}{=} a \int x \, dt + b \int y \, dt$$

$$a \int x \, dt + b \int y \, dt \stackrel{?}{=} a \int x \, dt + b \int y \, dt \qquad \text{This is an identity.}$$

So, we have shown that the differential operator is a linear operator. What about the square operator, $L[x] = x^2$?

$$[ax + by]^2 \stackrel{?}{=} ax^2 + by^2$$

$$a^2 x^2 + 2abxy + b^2 y^2 \stackrel{?}{=} ax^2 + by^2$$

$$a(1-a)x^2 + 2abxy + b(1-b)y^2 \stackrel{?}{=} 0$$

we can use the quadratic equation to solve for x:

$$x = \frac{-ab \pm \sqrt{a^2 b^2 - a(1-a)b(1-b)y^2}}{a(1-a)}$$

For any given value of y, the solution to this quadratic formula are the only solutions which satisfy equation (3.12). In order for the operator to be linear, equation (3.12) must be satisfied for all x. Therefore, the square operator is not a linear operator.

Now, let's see if the mean is a linear operator (we will do this just for the continuous case, but the result could also be shown for the discrete case):

$$E[ax + by] \stackrel{?}{=} aE[x] + bE[y]$$

Substitute in the definition of the mean from equation (3.1)

$$\int_{-\infty}^{\infty} [ax + by] f(x) \, dx \stackrel{?}{=} a \int_{-\infty}^{\infty} x f(x) \, dx + b \int_{-\infty}^{\infty} y f(x) \, dx$$

The integral of a sum is the sum of the integrals. Constants can be pulled outside the integral, so

$$a\int_{-\infty}^{\infty} xf(x)dx + b\int_{-\infty}^{\infty} yf(x)dx = a\int_{-\infty}^{\infty} xf(x)dx + b\int_{-\infty}^{\infty} yf(x)dx$$

This is an identity. The mean is a linear operator. As a result, we have a few simplifications for the mean. In the equations below, we assume that a and b are constants.

The mean of a constant is the constant.

$$E(a) = a$$

Adding a constant to a random variable adds the same constant to the mean.

$$E(ax + b) = aE(x) + b$$

The mean of the sum is the sum of the means.

$$E(g(x) + h(x)) = E(g(x)) + E(h(x))$$

These rules also apply to joint PDFs.

$$E(g(x, y) \pm h(x, y)) = E(g(x, y)) + E(h(x, y))$$

If and only if x and y are **independent** random variables, then

$$E(xy) = E(x)E(y) \qquad \text{(only for independent } x \text{ and } y\text{)}$$

We can show that the variance is **not** a linear operator. However, by substituting in for the definition of the variance, equation (3.4), we can come up with several short-cuts for computing some variances of functions. Again, we assume that a and b are constants.

The variance of a constant is zero.

$$\sigma_b^2 = 0$$

Adding a constant to a random variable does not change the variance of the random variable.

$$\sigma_{ax+b}^2 = a^2 \sigma_x^2$$

The variance of the sum is not the sum of the variances.

Expectations - 63

$$\sigma^2_{ax+by} = a^2\sigma^2_x + b^2\sigma^2_y + 2ab\sigma_{xy}$$

If and only if x and y are **independent**, then

$$\sigma^2_{ax+by} = a^2\sigma^2_x + b^2\sigma^2_y \qquad \text{(only for independent } x \text{ and } y\text{)}$$

We did not just make any of these theorems up. They can all be derived. As an example we now derive,

$$\sigma^2_{ax+by} = a^2\sigma^2_x + b^2\sigma^2_y + 2ab\sigma_{xy}$$

We begin by direct substitution of $(ax+by)$ into the definition of the variance:

$$\sigma^2_{g(x,y)} \equiv E\left[(g(x,y) - \mu_{g(x,y)})^2\right] = \int_{-\infty}^{\infty}\int_{-\infty}^{\infty}(g(x,y) - \mu_{g(x,y)})^2 f(x,y)dxdy$$

$$\sigma^2_{g(x,y)} = \int_{-\infty}^{\infty}\int_{-\infty}^{\infty}(ax+by-\mu_{ax+by})^2 f(x,y)dxdy$$

$$\sigma^2_{g(x,y)} = \int_{-\infty}^{\infty}\int_{-\infty}^{\infty}(a^2x^2 + b^2y^2 + 2abxy - 2ax\mu_{ax+by} - 2by\mu_{ax+by} + \mu_{ax+by}^2)f(x,y)dxdy$$

$$\sigma^2_{g(x,y)} = \int_{-\infty}^{\infty}\int_{-\infty}^{\infty}a^2x^2 f(x,y)dxdy + \int_{-\infty}^{\infty}\int_{-\infty}^{\infty}b^2y^2 f(x,y)dxdy + \int_{-\infty}^{\infty}\int_{-\infty}^{\infty}2abxyf(x,y)dxdy$$

$$- \int_{-\infty}^{\infty}\int_{-\infty}^{\infty}2ax\mu_{ax+by}f(x,y)dxdy - \int_{-\infty}^{\infty}\int_{-\infty}^{\infty}2by\mu_{ax+by}f(x,y)dxdy + \int_{-\infty}^{\infty}\int_{-\infty}^{\infty}\mu_{ax+by}^2 f(x,y)dxdy$$

$$\sigma^2_{g(x,y)} = a^2\int_{-\infty}^{\infty}\int_{-\infty}^{\infty}x^2 f(x,y)dxdy + b^2\int_{-\infty}^{\infty}\int_{-\infty}^{\infty}y^2 f(x,y)dxdy + 2ab\int_{-\infty}^{\infty}\int_{-\infty}^{\infty}xyf(x,y)dxdy$$

$$- 2a\mu_{ax+by}\int_{-\infty}^{\infty}\int_{-\infty}^{\infty}xf(x,y)dxdy - 2b\mu_{ax+by}\int_{-\infty}^{\infty}\int_{-\infty}^{\infty}yf(x,y)dxdy + \mu_{ax+by}^2 \int_{-\infty}^{\infty}\int_{-\infty}^{\infty}f(x,y)dxdy$$

$$\sigma^2_{g(x,y)} = a^2\mu_{x^2} + b^2\mu_{y^2} + 2ab\mu_{xy} - 2a\mu_{ax+by}\mu_x - 2b\mu_{ax+by}\mu_y + \mu_{ax+by}^2$$

$$\mu_{ax+by} = a\mu_x + b\mu_y$$

$$\sigma^2_{g(x,y)} = a^2\mu_{x^2} + b^2\mu_{y^2} + 2ab\mu_{xy} - 2a^2\mu_x^2 - 2ab\mu_y\mu_x - 2b^2\mu_y^2 - 2ab\mu_y\mu_x$$
$$+ a^2\mu_x^2 + b^2\mu_y^2 + 2ab\mu_x\mu_y$$

$$\sigma^2_{g(x,y)} = a^2\left(\mu_{x^2} - \mu_x^2\right) + b^2\left(\mu_{y^2} - \mu_y^2\right) + 2ab\left(\mu_{xy} - \mu_y\mu_x\right)$$

$$\sigma^2_{g(x,y)} = a^2\sigma^2_x + b^2\sigma^2_y + 2ab\sigma_{xy}$$
Q.E.D.

3.11. An Extended Example for Discrete Random Variables

Consider the isomerization reaction:

$$A \rightarrow B$$

This reaction takes place in a plant which relies on raw material solution, which unfortunately, is supposed to have a concentration of reactant of 1.0 mol/liter but in reality varies +/- 20%. The reactor is jacketed and is supposed to be isothermal. Day to day observation of the thermocouples in the reactor indicates that temperature swings about 10% around its set point of 300 K.

The reaction rate is given as

$$r_b = kC_A = k_o e^{-\frac{E_a}{RT}} C_A$$

where k is the rate constant, k_o is the pre-exponential factor of the rate constant, E_a is the activation energy, R is the gas constant, T is the temperature, and C_A is the concentration of the reactant. In one such reaction, $k_o = 20 \frac{liters}{min}$, $E_a = 10 \frac{kJ}{mol}$, and $R = 8.314 \frac{J}{mol \cdot K}$. Over a month, 20 spot measurements are made of the reactor, measuring the concentration of the reactant and the temperature.

Consider that the probability of obtaining any of the data points was uniform. Therefore, $f(x) = \frac{1}{n}$ where n is the number of measurements taken.

The tabulated data and the functions of that data are given in Table 3.3.

We use the definition of the mean, $\mu = E(x) = \sum_x xf(x)$, to obtain expectation values for the following functions: C_A, T, r_B, C_A^2, T^2, r_B^2, $C_A \cdot T$, $C_A \cdot r_B$ and $T \cdot r_B$. The expectations are shown in the table above in the row marked mean. The variances of C_A, T, and r_B are calculated using the "difference between the mean of the square and the square of the mean" rule.

$$\sigma^2_{g(x)} = E[g(x)^2] - E[g(x)]^2$$

Those variances are shown in the first three columns in the row marked variance. The covariances are obtained using the formula:

$$\sigma_{xy} = E[xy] - E[x]E[y]$$

and are shown in the last three columns for $C_A \cdot T$, $C_A \cdot r_B$ and $T \cdot r_B$. The standard deviations and correlation coefficients are given in the bottom row, obtained from:

$$\sigma_{g(x)} = \sqrt{\sigma^2_{g(x)}} \text{ and } \rho_{xy} = \frac{\sigma_{xy}}{\sigma_x \sigma_y}.$$

runs	C_A	T	r_B	C_A^2	T^2	r_B^2	$C_A \cdot T$	$C_A \cdot r_B$	$T \cdot r_B$
	mol/liter	K	mol/min						
1	1.11	296.49	0.38	1.22	87904.81	0.15	327.92	0.42	113.49
2	1.01	272.80	0.25	1.02	74419.84	0.06	275.60	0.25	67.06
3	1.03	270.22	0.24	1.06	73020.02	0.06	278.42	0.25	64.96
4	0.82	324.55	0.40	0.67	105332.35	0.16	265.94	0.33	130.71
5	0.83	273.87	0.21	0.70	75006.19	0.04	228.67	0.17	56.61
6	1.11	274.20	0.28	1.23	75185.95	0.08	304.31	0.31	75.73
7	0.80	299.56	0.29	0.64	89733.67	0.08	239.93	0.23	86.56
8	0.84	325.13	0.42	0.71	105709.64	0.17	273.64	0.35	135.39
9	0.89	310.19	0.37	0.78	96220.00	0.13	274.75	0.32	113.75
10	1.16	271.78	0.28	1.35	73862.15	0.08	315.23	0.32	75.44
11	1.13	298.13	0.40	1.27	88878.56	0.16	336.10	0.45	118.94
12	1.14	304.56	0.44	1.30	92759.55	0.19	347.10	0.50	133.77
13	1.13	270.54	0.26	1.27	73189.31	0.07	305.20	0.30	71.58
14	1.04	280.21	0.29	1.09	78514.87	0.08	292.34	0.30	79.93
15	1.15	306.72	0.46	1.32	94077.94	0.21	352.47	0.52	139.67
16	0.87	319.16	0.40	0.76	101864.25	0.16	277.89	0.35	128.30
17	0.83	304.60	0.32	0.68	92778.97	0.10	251.76	0.26	97.07
18	1.06	303.42	0.40	1.11	92064.18	0.16	320.26	0.42	121.60
19	0.83	289.58	0.26	0.69	83856.40	0.07	241.11	0.22	75.75
20	0.89	301.14	0.33	0.79	90684.97	0.11	267.54	0.29	98.58
sum	19.66	5896.84	6.66	19.68	1745063.64	2.32	5776.21	6.57	1984.89
mean	0.98	294.84	0.33	0.98	87253.18	0.12	288.81	0.33	99.24
variance	0.02	321.47	0.01		covariance		-1.03	0.00	1.09
standard deviation	0.13	17.93	0.07		correlation		-0.43	0.14	0.83

Table 3.3. Sample data and calculations for isomerization example of Section 3.11.

There is a sensible physical explanation of the expectations generated above. The mean and the standard deviation of the concentration show that statistically speaking:

$$C_A = 0.98 \pm 0.13 \frac{mol}{liter}$$

Similarly, $T = 294.5 \pm 17.9 K$ and $r_B = 0.33 \pm 0.07 \frac{mol}{min}$.

The physical meaning of the correlation coefficients are as follows. The concentration of A and the temperature (two independent random variables) should not be correlated. The correlation coefficient should be zero. It is -0.43. This non-zero value is a result of only having 20 data points. More data points would eventually average out to a correlation coefficient of zero.

The $C_A \cdot r_B$ correlation coefficient should be positive because as the concentration increases, the reaction rate increases. It is positive. The $C_A \cdot r_B$ correlation coefficient is small because the relationship is a linear (weak) relationship.

The $T \cdot r_B$ correlation coefficient should be positive because as the temperature increases, the reaction rate increases. It is positive. The $T \cdot r_B$ correlation coefficient is large because the relationship is an exponential (strong) relationship.

3.12. An Extended Example for Continuous Random Variables

A construction company has designed a distribution function which describes the area of their construction sites. The sites are all rectangular with dimensions a and b. The Joint PDF of the dimensions a and b are:

$$f(a,b) = \begin{cases} \frac{4}{21} ab & \text{for } 1 \leq a < 2 \text{ and } 3 \leq b \leq 4 \\ 0 & \text{otherwise} \end{cases}$$

The company is interested in determining pre-construction site costs including fencing and clearing land. The amount of fencing gives rise to a perimeter cost. The Perimeter Costs, PC, are $10 per meter of fencing required:

$$PC(a,b) = 10(2a + 2b)$$

The amount of land cleared is proportional to the area of the site and gives rise to an area cost. The Area Costs, AC, are $20 per square meter of the lot:

$$AC(a,b) = 20ab$$

(a) Are a and b independent?
(b) Find the mean of a, b, PC, and AC.
(c) Find the variance of a, b, PC, and AC.

Expectations - 67

(d) Find the covariance of $a \cdot b$, $a \cdot PC$, $a \cdot AC$, $b \cdot PC$, $b \cdot AC$, and $PC \cdot AC$.

(e) Find the correlation coefficient of $a \cdot b$, $a \cdot PC$, $a \cdot AC$, $b \cdot PC$, $b \cdot AC$, and $PC \cdot AC$.

(a) a and b are independent if $f(x,y) = g(x)h(y)$ where the marginal distributions are defined in Chapter 2 as

$$g(x) = \int_{-\infty}^{\infty} f(x,y)dy \quad \text{and} \quad h(y) = \int_{-\infty}^{\infty} f(x,y)dx$$

We evaluate the marginal distributions.

$$g(a) = \int_3^4 f(a,b)db = \int_3^4 \frac{4}{21}abdb = \frac{4}{21}a\frac{b^2}{2}\bigg|_3^4 = \frac{2}{3}a$$

$$h(b) = \int_1^2 f(a,b)da = \int_1^2 \frac{4}{21}abda = \frac{4}{21}b\frac{a^2}{2}\bigg|_1^2 = \frac{2}{7}b$$

$$f(x,y) = \frac{4}{21}ab = g(x)h(y) = \left(\frac{2}{3}a\right)\left(\frac{2}{7}b\right) = \frac{4}{21}ab$$

Therefore, a and b are independent.

(b) Find the mean of a, b, PC, and AC.

The general formula for the mean is:

$$\mu_{h(x,y)} = E(h(X,Y)) = \int_{-\infty}^{\infty}\int_{-\infty}^{\infty} h(x,y)f(x,y)dxdy$$

$$\mu_a = E(a) = \int_1^2 \int_3^4 a\left(\frac{4}{21}ab\right)dbda = \int_1^2 \left(\frac{4}{21}a^2\frac{b^2}{2}\right)_3^4 da = \left(\frac{4}{21}\frac{a^3}{3}\frac{7}{2}\right)_1^2 = \frac{98}{63} = 1.56\,\text{m}$$

$$\mu_b = E(b) = \int_1^2 \int_3^4 b\left(\frac{4}{21}ab\right)dbda = \int_1^2 \left(\frac{4}{21}a\frac{b^3}{3}\right)_3^4 da = \left(\frac{4}{21}\frac{a^2}{2}\frac{37}{3}\right)_1^2 = \frac{74}{21} = 3.52\,\text{m}$$

We can use the definition of the mean to compute the mean value of the perimeter costs.

$$\mu_{PC} = E(PC) = \int_1^2 \int_3^4 20(a+b)\left(\frac{4}{21}ab\right)dbda = \frac{80}{21}\int_1^2 \left(a^2\frac{b^2}{2} + a\frac{b^3}{3}\right)_3^4 da =$$

$$= \frac{80}{21}\left(\frac{a^3}{3}\frac{7}{2} + \frac{a^2}{2}\frac{37}{3}\right)_1^2 = \frac{80}{21}\left(\frac{49}{6} + \frac{111}{6}\right) = \frac{6400}{63} = \$101.59$$

OR remember that the mean is a linear operator, $E(ax+by) = aE(x) + bE(y)$

$$\mu_{PC} = E(PC) = E(20a) + E(20b) = 20E(a) + 20E(b) = 20(1.56) + 20(3.52) = \$101.6$$

For the area cost, we can use the definition of the mean.

$$\mu_{AC} = E(AC) = \int_1^2 \int_3^4 20(ab)\left(\frac{4}{21}ab\right)dbda = \frac{80}{21}\int_1^2 \left(a^2\frac{b^3}{3}\right)_3^4 da = \frac{80}{21}\left(\frac{a^3}{3}\frac{37}{3}\right)_1^2 = \frac{20720}{189} = \$109.63$$

OR remember $E(xy) = E(x)E(y)$ for independent variables.

$$\mu_{AC} = E(AC) = E(20ab) \ne 20E(a)E(b) = 20(1.56)(3.52) = \$109.63$$

because a and b are independent.

(c) Find the variance of a, b, PC, and AC.

The working equation to calculate the variance of a function is:

$$\sigma^2_{g(x,y)} = E[g(x,y)^2] - E[g(x,y)]^2$$

For these variables, we have calculated the mean (necessary to evaluate the function in the second term on the right hand side). We must next calculate the mean of the square (the first term on the right hand side) before we can calculate the variance.

$$\mu_{a^2} = E(a^2) = \int_1^2 \int_3^4 a^2 \left(\frac{4}{21}ab\right)dbda = \int_1^2 \left(\frac{4}{21}a^3\frac{b^2}{2}\right)_3^4 da = \left(\frac{4}{21}\frac{a^4}{4}\frac{7}{2}\right)_1^2 = \frac{420}{168} = 2.50$$

$$\sigma^2_a = E[a^2] - E[a]^2 = 2.50 - 1.5556^2 = 0.0802$$

Expectations - 69

$$\mu_{b^2} = E(b^2) = \int_1^2\int_3^4 b^2\left(\frac{4}{21}ab\right)dbda = \int_1^2\left(\frac{4}{21}a\frac{b^4}{4}\right)_3^4 da = \left(\frac{4}{21}\frac{a^2}{2}\frac{175}{4}\right)_1^2 = \frac{525}{42} = 12.50$$

$$\sigma_b^2 = E[b^2] - E[b]^2 = 12.50 - 3.5238^2 = 0.0828$$

To calculate the variance of $PC(a,b) = 10(2a + 2b)$ remember

$$\sigma_{aX+bY}^2 = a^2\sigma_X^2 + b^2\sigma_Y^2 + 2ab\sigma_{XY}$$

Then we only need to calculate the covariance of a and b. The working formula for the covariance is:

$$\sigma_{XY} = E[xy] - E[x]E[y]$$

So we need the expectation value of $E(ab)$

$$\mu_{ab} = E(ab) = \int_1^2\int_3^4 (ab)\left(\frac{4}{21}ab\right)dbda = \frac{4}{21}\int_1^2\left(a^2\frac{b^3}{3}\right)_3^4 da = \frac{4}{21}\left(\frac{a^3}{3}\frac{37}{3}\right)_1^2 = \frac{1036}{189} = 5.48$$

Then

$$\sigma_{ab} = 5.48 - 1.56(3.52) = 0.00$$

The covariance of a and b is zero. We should have known that because we showed in part (a) that a and b were statistically independent.

$$\sigma_{PC}^2 = \sigma_{20(a+b)}^2 = 20^2(\sigma_a^2 + \sigma_b^2) + 2(20)(20)\sigma_{ab} = 65.2.$$

Lastly, for the area cost,

$$\mu_{AC^2} = E(AC^2) = \int_1^2\int_3^4 20^2(ab)^2\left(\frac{4}{21}ab\right)dbda = \frac{1600}{21}\int_1^2\left(a^3\frac{b^4}{4}\right)_3^4 da =$$

$$= \frac{1600}{21}\left(\frac{a^4}{4}\frac{175}{4}\right)_1^2 = \frac{4200000}{336} = 12500.0$$

$$\sigma_{AC}^2 = E[AC^2] - E[AC]^2 = 12500.0 - 109.63^2 = 481.$$

(d) Find the covariance of $a \cdot b$, $a \cdot PC$, $a \cdot AC$, $b \cdot PC$, $b \cdot AC$, and $PC \cdot AC$.

In part (c) we found the covariance of $a \cdot b$ to be 0.0 because they were statistically independent. For the rest of these quantities, we use the rule

$$\sigma_{XY} = E[xy] - E[x]E[y]$$

where we already have the expectation values of the two factors in the second term on the r.h.s. We only need to find the first term on the r.h.s. to find the covariance. For the covariance between a and the perimeter cost we have

$$\mu_{aPC} = E(aPC) = \int_1^2 \int_3^4 20(a^2 + ab)\left(\frac{4}{21}ab\right)dbda = \frac{80}{21}\int_1^2 \left(a^3 \frac{b^2}{2} + a^2 \frac{b^3}{3}\right)_3^4 da =$$

$$= \frac{80}{21}\left(\frac{a^4}{4}\frac{7}{2} + \frac{a^3}{3}\frac{37}{3}\right)_1^2 = \frac{80}{21}\left(\frac{105}{8} + \frac{259}{9}\right) = \frac{241360}{1512} = 159.63$$

$$\sigma_{aPC} = E[aPC] - E[a]E[PC] = 159.63 - (1.556)(101.59) = 1.56$$

For the covariance between b and the perimeter cost we have

$$\mu_{bPC} = E(bPC) = \int_1^2 \int_3^4 20(ab + b^2)\left(\frac{4}{21}ab\right)dbda = \frac{80}{21}\int_1^2 \left(a^2 \frac{b^3}{3} + a \frac{b^4}{4}\right)_3^4 da =$$

$$= \frac{80}{21}\left(\frac{a^3}{3}\frac{37}{3} + \frac{a^2}{2}\frac{175}{4}\right)_1^2 = \frac{80}{21}\left(\frac{259}{9} + \frac{525}{8}\right) = \frac{543760}{1512} = 359.63$$

$$\sigma_{bPC} = E[bPC] - E[b]E[PC] = 359.63 - (3.5238)(101.59) = 1.65$$

For the covariance between a and the area cost we have

$$\mu_{aAC} = E(aAC) = \int_1^2 \int_3^4 20(a^2 b)\left(\frac{4}{21}ab\right)dbda = \frac{80}{21}\int_1^2 \left(a^3 \frac{b^3}{3}\right)_3^4 da =$$

$$= \frac{80}{21}\left(\frac{a^4}{4}\frac{37}{3}\right)_1^2 = \frac{44400}{252} = 176.19$$

$$\sigma_{aAC} = E[aAC] - E[a]E[AC] = 176.19 - (1.556)(109.63) = 5.65$$

Expectations - 71

For the covariance between b and the area cost we have

$$\mu_{bAC} = E(bAC) = \int_1^2 \int_3^4 20(ab^2)\left(\frac{4}{21}ab\right)dbda = \frac{80}{21}\int_1^2 \left(a^2 \frac{b^4}{4}\right)_3^4 da =$$

$$= \frac{80}{21}\left(\frac{a^3}{3} \frac{175}{4}\right)_1^2 = \frac{98000}{252} = 388.89$$

$$\sigma_{bAC} = E[bAC] - E[b]E[AC] = 388.89 - (3.52)(109.63) = 3.00$$

For the covariance between the perimeter cost and the area cost we have

$$\mu_{ACPC} = E(ACPC) = \int_1^2 \int_3^4 20^2(a^2b + ab^2)\left(\frac{4}{21}ab\right)dbda = \frac{1600}{21}\int_1^2 \left(a^3 \frac{b^3}{3} + a^2 \frac{b^4}{4}\right)_3^4 da =$$

$$= \frac{1600}{21}\left(\frac{a^4}{4} \frac{37}{3} + \frac{a^3}{3} \frac{175}{4}\right)_1^2 = \frac{1600}{21}\left(\frac{555}{12} + \frac{1225}{12}\right) = \frac{2848000}{252} = 11301.63$$

$$\sigma_{ACPC} = E[ACPC] - E[AC]E[PC] = 11301.63 - (109.63)(101.59) = 164.3$$

(e) Find the correlation coefficient of $a \cdot b$, $a \cdot PC$, $a \cdot AC$, $b \cdot PC$, $b \cdot AC$, and $PC \cdot AC$.

The general formula for the correlation coefficient is:

$$\rho_{XY} = \frac{\sigma_{XY}}{\sigma_X \sigma_Y}$$

$$\rho_{ab} = \frac{\sigma_{ab}}{\sigma_a \sigma_b} = \frac{0.0}{\sqrt{1.56}\sqrt{3.52}} = 0.0$$

$$\rho_{PCAC} = \frac{\sigma_{PCAC}}{\sigma_{PC}\sigma_{AC}} = \frac{164.3}{\sqrt{65.2}\sqrt{481}} = 0.93$$

$$\rho_{aPC} = \frac{\sigma_{aPC}}{\sigma_a \sigma_{PC}} = \frac{1.56}{\sqrt{1.56}\sqrt{65.2}} = 0.15$$

$$\rho_{bPC} = \frac{\sigma_{bPC}}{\sigma_b \sigma_{PC}} = \frac{1.65}{\sqrt{3.52}\sqrt{65.2}} = 0.11$$

$$\rho_{aAC} = \frac{\sigma_{aAC}}{\sigma_a \sigma_{AC}} = \frac{5.65}{\sqrt{1.56}\sqrt{481.}} = 0.21$$

$$\rho_{bAC} = \frac{\sigma_{bAC}}{\sigma_b \sigma_{AC}} = \frac{3.00}{\sqrt{3.52}\sqrt{481}} = 0.07$$

These correlations (with the exception of a and b) are all positive. They should be because as you increase one side of the lot (either a or b), you should increase both the perimeter and the area. Also, as you increase the perimeter, on average, you increase the area, given our distribution function.

3.13. Subroutines

Code 3.1. Variance as a function of truncation (vartrunc.m)

This simple code illustrates the need to keep all significant figures in the intermediate calculations of averages before computing the variance using equation (3.5).

```
n=10;
r = rand(n,1);
s = 10 + 0.1*(2*r - 1)
s2 = s.^2;
f = 1/n;
format long
mu_s = sum(f*s);
mu_s2 = sum(f*s2);
var_s = mu_s2 - mu_s^2;
for i = 2:1:8
   mu_s_cut(i) = round(mu_s*(10^i))/(10^i);
   mu_s2_cut(i) = round(mu_s2*(10^i))/(10^i);
   var_s_cut(i) = mu_s2_cut(i) - mu_s_cut(i)^2;
   fprintf(1,'%i sig figs:   mu_s = %16.9e mu_s2 = %16.9e var_s = %16.9e\n',
i+2, mu_s_cut(i),mu_s2_cut(i),var_s_cut(i));
end
   fprintf(1,'all sig figs: mu_s = %16.9e mu_s2 = %16.9e var_s = %16.9e\n',
mu_s,mu_s2,var_s);
```

3.14. Problems

Problem 3.1.

A chemical plant produces A thousands of liters of "A-plus Liquid Fungicide" and B thousands of liters of "B-Gone Liquid Insecticide" per month. The two processes share some raw materials and facilities so that the amount of A and B produced are not independent of each other. In fact the amount of B produced is related to the amount of A produced by

$$B = RM - \frac{A}{2} + 40$$

where RM is the amount of raw materials received at the plant in a given month (also in liters). The total amount of product in thousands of liters is given as

$$T(A, B) = B + A$$

The monthly production schedule for 2012 is as followed

Month	A (thousands of liters)	RM (thousands of liters)
Jan	50	120
Feb	50	120
Mar	60	110
Apr	70	120
May	80	130
Jun	90	130
Jul	100	130
Aug	100	130
Sep	90	130
Oct	80	120
Nov	70	110
Dec	60	100

In all problems, remember to put units with your answers. In all relevant problems, you are encouraged to write down the appropriate formula before you use it.
  (a) Is this problem continuous or discrete?
  (b) Find the average monthly production of A.
  (c) Find the average monthly production of B. B is just a function of A and RM.
  (d) Find the average monthly usage of RM.
  (e) Find the mean of the total monthly production, T.
  (f.1) Find the variance of the monthly production of A using the rigorous definition of the variance.
  (f.2) Find the variance of the monthly production of A using the "mean of the squares minus the square of the mean" formula.
  (g) Find the variance of the monthly usage of RM.
  (h.1) Find the variance of the monthly production of B from tabulated values of B.
  (h.2) Find the variance of the monthly production of B from the variances of A and RM and the formula for B given in this problem statement.
  (i) Find the variance of the total monthly production, T.
  (j,k,l,m) Find the standard deviations of A, B, RM and T.

(n) Find the covariance of A and B.
(o) Find the correlation coefficient of A and B
(p) Give a physical description of what the value and sign of the correlation coefficient means.

Problem 3.2.

A chemical plant contains a jacketed vessel in which the following isomerization reaction takes place:

$$A \rightarrow B$$

The rate of the production of B, r_B [moles/hour], is given by

$$r_B = kC_A$$

where C_A is the concentration of A [moles/liter] and the reaction rate constant, k [liters/hour], is given as a function of the temperature, T [Kelvin], as

$$k = 20.0 \cdot e^{-\frac{10,000}{RT}}$$

where R is the gas constant [8.314 J/mole/K]. This (highly ideal) jacketed vessel keeps temperature perfectly constant at the set temperature of 400 K. The concentration in the tank is obtained from the mass balance

$$accumulation = in - out + generation$$

$$V \frac{dC_A}{dt} = QC_{A,in} - QC_A - kC_A$$

where Q is the volumetric flowrate [liters/hour], and has a numerical value of $Q = 9.0$ l/hour. V is the reactor volume, $V = 100.0$ liters. Rearrangement yields:

$$\left(\frac{V}{QC_{A,in} - QC_A - kC_A} \right) dC_A = dt$$

and where $C_{A,in}$ is the inlet concentration of A, $C_{A,in} = 2.0$ mole/liter. We can integrate this equation to yield

$$\frac{V}{(Q+k)} \ln\left(\frac{kC_{A,in}}{QC_{A,in} - QC_A - kC_A} \right) = t$$

We can rearrange this equation to give us C_A

$$C_A(t) = \frac{C_{A,in}}{(Q+k)}\left(Q + ke^{-\left(\frac{Q+k}{V}\right)t}\right)$$

(a) Plot C_A and C_B on one graph and plot r_B as functions of t for $0 \le t \le 24$ hour. Remember, $C_B = C_{A,in} - C_A$.

(b) For our problem at hand, identify x, a, b, $h(x)$, and $f(x)$.
(c) What is the average concentration of reactant, C_A, during that first day of operation?
(d) What is the average rate of production, r_B, during that first day of operation?
(e) What is the average concentration of B, C_B, during that first day of operation?
(f) What is the variance of C_A during that first day of operation?
(g) What is the variance of r_B during that first day of operation?
(h) What is the variance of C_B during that first day of operation?

Problem 3.3.
A private pilot wishes to insure his airplane for $1,000,000. The insurance company estimates that a total loss may occur with a probability of 0.001, a 50% loss with probability 0.01 and a 25% loss with a probability of 0.1. Ignoring all other partial losses, what premium should the insurance company charge each year to realize an average annual profit of $1000 on this policy?

Chapter 4. Discrete Probability Distributions

4.1. Introduction

In Chapter 2, we learned how to compute probabilities and cumulative probabilities for arbitrary discrete and continuous probability distribution functions (PDFs). In Chapter 3, we learned how to compute expectations, such as the mean and variance, of random variables and functions of them, given a PDF, $f(x)$. In this chapter, we introduce some of the commonly occurring PDFs for discrete sample spaces. We will apply the rules of Chapters 2 and 3, to use the PDFs to calculate probabilities and expectations. The goal of this section is to become familiar with these probability distributions and, when given a problem, know which PDF is appropriate.

4.2. Discrete Uniform Distribution

If the random variable X assumes the values of $x_1, x_2, x_3 \ldots x_k$ with equal probability, then the discrete uniform distribution is given by $f(x;k)$ (The semicolon is used to separate random variables, which shall always appear before the semicolon, from parameters, which appear after.)

$$f(x;k) = \frac{1}{k} \tag{4.1}$$

In calculating the mean and variance of the Discrete Uniform Distribution PDF, or any discrete PDF for that matter, we have a definition given in equation (3.1.a), namely

$$\mu_x = E(x) = \sum_x x f(x) \tag{3.1.a}$$

Using this formula will always give the correct result. However, if the number of elements in the sample space is infinite, it will not be practical to explicitly evaluate each term in the summation. For many of these commonly occurring PDFs, the evaluation of the definition of the mean reduces to a formula for the mean. In some cases, the formula is easy to derive and in other cases, more difficult. Some PDFs have no simple expression for the mean or variance.

For the Discrete Uniform Distribution, the PDF is a constant. Consequently, the mean of the random variable, x, is given by

$$\mu_x = \frac{1}{k}\sum_{i=1}^{k} x_i \qquad (4.2)$$

The variance of the random variable, x, obeying the discrete uniform distribution is given by

$$\sigma_x^2 = \frac{1}{k}\sum_{i=1}^{k}(x_i - \mu_x)^2 \qquad (4.3)$$

An example of a histogram of the discrete uniform distribution is shown in Figure 4.1.

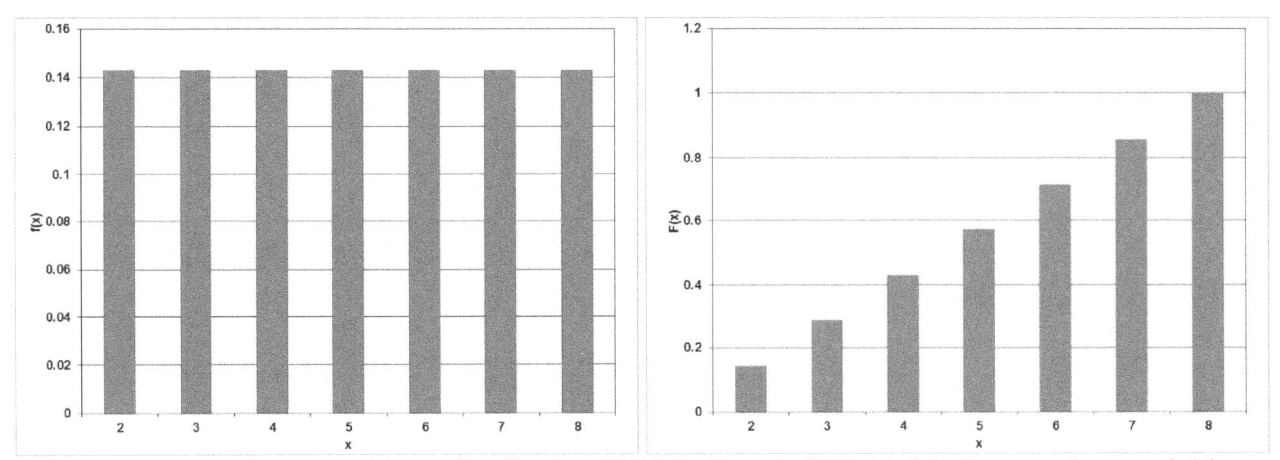

Figure 4.1. The discrete uniform distribution PDF (left) and CDF (right) for a random variable, x, taking integer values between 2 and 8 inclusive.

Example 4.1.: You select a card randomly from a mixed deck of cards. What is the probability you draw a king with an axe or a one-eyed jack? (Note, there is only one king with an axe but there are two jacks shown in profile.) If you assign a numeric value of 1 to the ace, 11 to the jack, 12 to the queen, and 13 to the king, what is the mean value of the card drawn? What is the variance?

The probability of drawing a king with an axe or a one-eyed jack is 3/52 by equation (4.1) and the union rule for mutually exclusive events, equation (1.34). The mean is 7 by equation (4.2). The variance is 14 by equation (4.3).

4.3. Relationship between binomial, multinomial, hypergeometric, and multivariate hypergeometric PDFs

The next four PDFs we are going to discuss are the binomial, multinomial, hypergeometric, and multivariate hypergeometric PDFs. Which of the four PDFs you need to employ for a given problem depends upon two criteria: (1) how many outcomes an experiment can yield, and (2) whether the probability of a particular outcome changes from one trial to the next. Frequently, the change in probability is due to not replacing some element of the experiment. Therefore, this second factor is noted as replacement vs. no replacement. Table 4.1 describes when each of the PDFs should be used.

	replacement	no replacement
2 outcomes	binomial	hypergeometric
n>2 outcomes	multinomial	multivariate hypergeometric

Table 4.1. Relationship between binomial, multinomial, hypergeometric, and multivariate hypergeometric PDFs.

4.4. Binomial Distribution

The binomial and multinomial distributions arise from a process called the Bernoulli Process. The Bernoulli process satisfies the following three criteria.

1. An experiment consists of n repeated, **independent** trials.
2. Each trial can have one of two outcomes, success or failure.
3. The probability of success, p, is the same for each trial.

Two common examples of Bernoulli processes are given below.

Example 4.2.: Flipping a coin n times is a Bernoulli process. A success is defined as landing heads up. Each toss is a trial. Each toss is independent. Each toss has one of two outcomes: heads or tails. Finally, the probability for heads is the same for each toss.

Example 4.3.: Grabbing a handful of marbles from a bag or red and black marbles, and **replacing** the marbles between grabs is a Bernoulli process. A success can be defined as more than m red marbles in hand. Each grab is a trial. Each grab is independent, so long as there is replacement. Each grab has one of two outcomes: more than m red marbles or less than or equal to m red marbles; success or failure. Sure the number of red marbles varies, but that's not our criterion for success, only more or less than m. Finally, the probability for success is the same for each grab.

The random variable, X, in a binomial distribution, $b(x; n, p)$, is the number of successes from n Bernoulli trials. So for our first example, flipping a coin n times, the probability of a getting a head in one independent trial is p. For n trials, the binomial random variable can assume values between 0 (never getting a head) up to n (getting a head every time). The distribution gives the probability for getting a particular value of successes in n trials.

The binomial distribution is (where q the probability of a failure is $q = 1 - p$)

$$P(X = x) = b(x; n, p) = \binom{n}{x} p^x q^{n-x} \tag{4.4}$$

Without derivation, the mean of the random variable, x, obeying the binomial PDF is

$$\mu_x = np \tag{4.5}$$

The variance of the random variable, x, obeying the binomial PDF is

$$\sigma_x^2 = npq \tag{4.6}$$

Frequently, we are interested in the cumulative PDF, as defined in equation (2.3).

$$P(X \leq r) = \sum_{i=1}^{x_i \leq r} f(x_i) \tag{2.3}$$

The cumulative probability distribution of the binomial PDF is obtained by substituting the binomial PDF in equation (4.) into the equation above,

$$P(X \leq r) \equiv B(r; n, p) = \sum_{x=0}^{r} b(x; n, p) \tag{4.7}$$

There are a variety of ways to calculate the cumulative binomial PDF for given values of r, n and p. In the old days, when cavemen wanted to calculate cumulative probabilities based on the binomial distribution, they turned to tables of values chiseled on the stone walls of their caves. In later years, these tables were transcribed into the appendices of statistics textbooks. Yet later, these same tables were transcribed onto files available on the internet, where a cursory search of "cumulative binomial distribution tables" will turn up numerous examples.

The disadvantage of using tables, aside from the fact that many of us may find the comparison to cavemen to be an unflattering one, is that the parameter, p, can take on any value between 0 and 1, but the tables only provide values of the cumulative PDF for a few values of p, usually, 0.1, 0.2...0.8, 0.9. For any other value of p, we must interpolate in the table, which adds an unnecessary error.

We can also use modern computational tools to evaluate the binomial PDF and the cumulative binomial PDF (the binomial CDF) for any arbitrary values of r, n and p. What follows are two paths to computing the PDF and CDF for the binomial distribution. In the first path, we write our own little codes. This exercise is instructive because it illustrates the simplicity of the process. In the second path, we use pre-existing routines available in MATLAB. This path is useful because it acknowledges the fact that many of the problems that face us have already been solved and robust codes exist and are at our disposal.

We can compute the binomial PDF, given in equation (4.4) with the following MATLAB code, binomial.m:

```
function f = binomial(x,n,p)
f = comb(n,x)*p^x*(1-p)^(n-x);
```

This two line program accepts as inputs x, n, and p, and returns $f = b(x;n,p)$. This code accesses the program, comb.m, to obtain the combinations. The code for comb.m is given as Code 1.3. in Chapter 1.

If we wanted the cumulative binomial PDF, $P(X \le r) \equiv B(r;n,p) = \sum_{x=0}^{r} b(x;n,p)$, then we could write a short code and call it, binocumu.m, which would contain

```
function f = binocumu(r,n,p)
f = 0.0
for x = 0:1:r
   f = f + binomial(x,n,p);
end
```

If, instead, we wanted the most general code to calculate the probability from the binomial PDF in some arbitrary interval, then we could write in the file binoprob.m

```
function f = binoprob(a,c,n,p)
f = 0.0
for x = a:1:c
   f = f + binomial(x,n,p);
end
```

This file returns the value of $P(a \le X \le c) = \sum_{x=a}^{c} b(x;n,p)$. In the table below we see how the program, binoprob.m can calculate the probability for any arbitrary interval, given the correct values of a and c. The table does not present a complete set of all the possible combinations but does give the general idea.

We note that in Table 4.2 any requested probability could be obtained with a single call to the function binoprob.m We will find shortly that it is useful to construct an analogous table, using only the cumulative distribution, binocumu.m.

probability	command line argument
$P(X = a) = P(a \leq X \leq a)$	`binoprob(a,a,n,p)`
$P(X \leq a) = P(0 \leq X \leq a)$	`binoprob(0,a,n,p)`
$P(X < a) = P(0 \leq X \leq a-1)$	`binoprob(0,a-1,n,p)`
$P(X \geq a) = P(a \leq X \leq n)$	`binoprob(a,n,n,p)`
$P(X > a) = P(a+1 \leq X \leq n)$	`binoprob(a+1,n,n,p)`
$P(a \leq X \leq c) = P(a \leq X \leq c)$	`binoprob(a,c,n,p)`
$P(a < X < c) = P(a+1 \leq X \leq c-1)$	`binoprob(a+1,c-1,n,p)`
$P(a \geq X \geq c) = P(c \leq X \leq a)$	`binoprob(c,a,n,p)`
$P(a > X > c) = P(c < X < a)$	`binoprob(c+1,a-1,n,p)`

Table 4.2. Using the code binoprob.m to compute various binomial probabilities.

probability	command line argument
$P(X = a) = P(a \leq X \leq a)$	`binocumu(a,n,p)-binocumu(a-1,n,p)`
$P(X \leq a) = P(0 \leq X \leq a)$	`binocumu(a,n,p)`
$P(X < a) = P(0 \leq X \leq a-1)$	`binocumu(a-1,n,p)`
$P(X \geq a) = P(a \leq X \leq n)$	`1-binocumu(a-1,n,p)`
$P(X > a) = P(a+1 \leq X \leq n)$	`1-binocumu(a,n,p)`
$P(a \leq X \leq c) = P(a \leq X \leq c)$	`binocumu(c,n,p)-binocumu(a-1,n,p)`
$P(a < X < c) = P(a+1 \leq X \leq c-1)$	`binocumu(c-1,n,p)-binocumu(a,n,p)`
$P(a \geq X \geq c) = P(c \leq X \leq a)$	`binocumu(a,n,p)-binocumu(c-1,n,p)`
$P(a > X > c) = P(c < X < a)$	`binocumu(a-1,n,p)-binocumu(c,n,p)`

Table 4.3. Using the code binocumu.m to compute various binomial probabilities.

We note that in Table 4.3 some probabilities require two calls to binocumu.m or that the result of that function be subtracted from 1. If the arguments of these calls (a or a-1 and c or c-1) seem confusing, draw out a sample histogram and label a and c. Mark the range that you want and then define it terms of cumulative PDFs.

We have shown above that the binomial PDF can be evaluated with very simple codes. MATLAB already has codes to evaluate the PDF and CDF. We now demonstrate how these codes can be used. A summary of these MATLAB commands is given in Appendix IV. In order to evaluate a PDF at a given value of x, one can use the `pdf` function in MATLAB

```
>> f = pdf('Binomial',x,n,p)
```

For example, to generate the value of the PDF for the binomial PDF defined for $x = 1$, $n = 4$ and $p = 0.33$, one can type the command,

```
>> f = pdf('Binomial',1,4,0.33)
```

which yields the following output (when the `format long` command has first been used to provide all sixteen digits)

```
f = 0.397007160000000
```

This tells us that the probability of getting 1 success in four attempts where the probability of success of an individual is given by 0.33 is about 0.397.

In order to evaluate a cumulative PDF or CDF at a given value of x, one can use the `cdf` function in MATLAB. If we are interested in the probability that $x \leq r$, then the appropriate function is the cumulative distribution function.

$$F = p(x \leq r) \equiv \sum_{x=0}^{r} b(x;n,p)$$

In MATLAB, we can directly evaluate the cumulative distribution function for a number of common PDFs.

```
>> F = cdf('Binomial',x,n,p)
```

For example, to generate the value of the binomial CDF for $x \leq 1$, $n = 4$ and $p = 0.33$, one can type the command,

```
>> F = cdf('Binomial',1,4,0.33)
```

which yields the following output

```
F = 0.598518370000000
```

An example of a histogram of the binomial distribution is shown in Figure 4.2.

Discrete Probability Distributions - 83

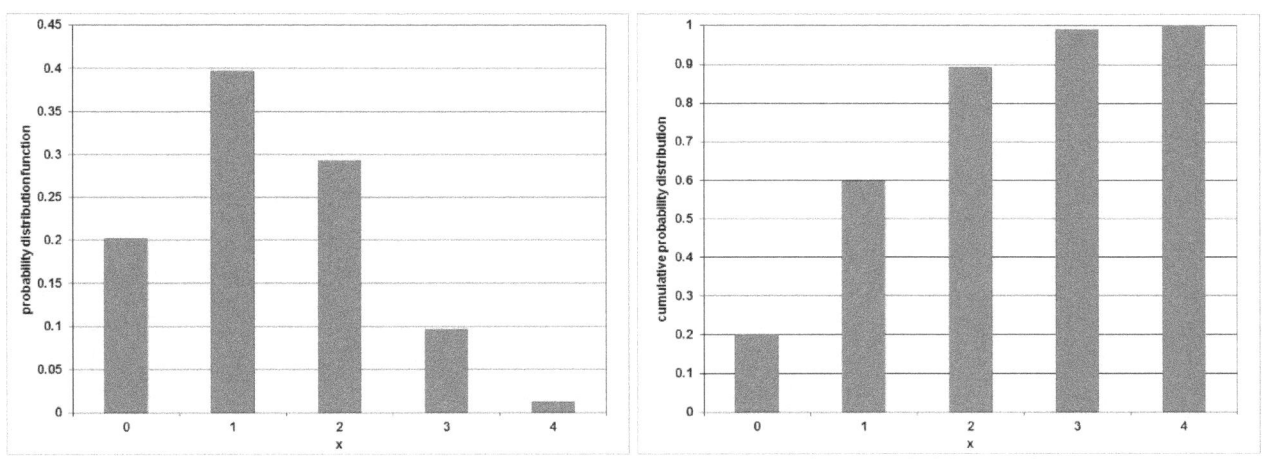

Figure 4.2. The binomial PDF (left) and CDF (right) for a random variable, *x*, given parameters *n* = 4 and *p* = 0.33.

Example 4.2: Consider the following experiment. We flip a coin 20 times. A success is defined as heads. The probability of a success in a single trial is 0.5.
(a) What is the probability that 5 of the 20 tosses are heads?
(b) What is the average number of heads
(c) What is the variance of the number of heads?
(d) What is the probability of getting between 8 and 12 heads, inclusive?

(a) What is the probability that 5 of the 20 tosses are heads?

From equation (4.4)

$$P(X = 5) = b(5;20,0.5) = \binom{20}{5} 0.5^5 0.5^{15} = 0.014786$$

Using the pdf command in MATLAB, we have:

```
>> f = pdf('Binomial',5,20,0.5)

f = 0.014785766601562
```

(b) What is the average number of heads

From equation (4.5)

$$\mu_x = np = 20 \cdot 0.5 = 10$$

(c) What is the variance of the number of heads?

From equation (4.6)

$$\sigma_x^2 = npq = 20 \cdot 0.5 \cdot 0.5 = 5$$

(d) What is the probability of getting between 8 and 12 heads, inclusive?

If we rely on the cumulative distribution function, we can write this probability as

$$P(8 \leq X \leq 12) = P(X \leq 12) - P(X < 8)$$

$$P(8 \leq X \leq 12) = P(X \leq 12) - P(X \leq 7)$$

$$P(8 \leq X \leq 12) = B(r = 12; n = 20, p = 0.5) - B(r = 7; n = 20, p = 0.5)$$

We can use a table from the web and obtain

$$P(8 \leq X \leq 12) = 0.8684 - 0.1316 = 0.7368$$

We can use the cdf function in MATLAB

```
>> p = cdf('Binomial',12,20,0.5) - cdf('Binomial',7,20,0.5)

F = 0.736824035644531
```

So when you give a coin 20 flips, roughly 74% of the time, you will wind up with between 8 and 12 heads, inclusive. Does this mean 30% of the time you will wind up with 8 to 12 tails, inclusive? Why or why not?

4.5. Multinomial Distribution

If a Bernoulli trial can have more than 2 outcomes (success or failure) then it ceases to be a Bernoulli trial and becomes a multinomial experiment. In the multinomial experiment, there are k outcomes and n trials. Each outcome has a result E_i. There are now k random variables X_i, each representing the probability of obtaining result E_i in X_i of the n trials.

The distribution of a multinomial experiment is

$$P(\underline{X} = \underline{x}) = m(\underline{x}; n, \underline{p}, k) = \binom{n}{x_1, x_2 \ldots x_k} \prod_{i=1}^{k} p_i^{x_i} \qquad (4.8)$$

where $\sum_{i=1}^{k} x_i = n$ and $\sum_{i=1}^{k} p_i = 1$.

Note that in order to evaluate the multinomial PDF, we require two vectors. A vector of probabilities for each outcome, $\underline{p}$, and a vector of the number of each outcome that we are interested in, $\underline{x}$. We can again use modern computational tools to evaluate the multinomial PDF for any arbitrary values of $\underline{x}$, n, $\underline{p}$ and k. What follows are two paths to computing the PDF for the multinomial distribution. In the first path, we write our own little code. This exercise is instructive because it illustrates the simplicity of the process. In the second path, we use a pre-existing routine available in MATLAB.

First, we can write a code to evaluate $P(\underline{X} = \underline{x}) = m(\underline{x}; n, \underline{p}, k)$, such as the one in multinomial.m

```
function prob = multinomial(x,n,p,k)
prob = factorial(n);
for i = 1:1:k
    prob = prob/factorial(x(i))*p(i)^x(i);
end
```

In the code, multinomial.m, x and p are vectors of length k. This code would be run at the command prompt with something like

```
>> f = multinomial([2,4,3],9,[0.5,0.3,0.2],3)
```

where $\underline{x} = [2,4,3]$, $n = 9$, $\underline{p} = [0.5,0.3,0.2]$ and $k = 3$. This command yields

```
f = 0.020412000000000
```

Alternatively, there is an intrinsic function in MATLAB, but it is not the usual pdf function. For the multinomial distribution, there is a special command, mnpdf.

```
>> f = mnpdf(x,p)
```

In this case, we didn't have to input the parameters n and k because the code knows that the sum of the elements in the vector x is n and the number of elements in the vector x is k. An example of the use of this function, where $\underline{x} = [2,4,3]$, $n = 9$, $\underline{p} = [0.5,0.3,0.2]$ and $k = 3$, is

```
>> f = mnpdf([2,4,3],[0.5,0.3,0.2])
```

This command yields

```
f = 0.020412000000000
```

There are not any intrinsic routines for computing probabilities over ranges of variables for the multinomial distribution. Simple codes could be written following the model of binoprob.m.

Example 4.3.: A hypothetical statistical analysis of people moving to Knoxville, TN shows that 25% of people who move here do so to attend the University, 55% move here for a professional position, and the remaining 20% for some other reason. If you ask 10 new arrivals in Knoxville, why they moved here, what is the probability that all of them moved here to go to UT?

In this problem, there are ten new arrivals. Each is considered an experiment, thus $n = 10$. There are three possible outcomes: U=university, P=profession and O=other, thus $k = 3$. The probabilities of each outcome are $p_U = 0.25$, $p_P = 0.55$ and $p_O = 0.20$. In this question we are asked to find the probability that $x_U = 10$, $x_P = 0$ and $x_O = 0$.

Using equation (4.7), we have

$$P(\{X\} = [10,0,0]) = m([10,0,0];10,[0.25,0.55,0.20],3) = \frac{10!}{10!0!0!} 0.25^{10} 0.55^0 0.20^0 = 9.54e\text{-}007$$

Using the mnpdf function in MATLAB, we can accomplish the same task.

```
>> f = mnpdf([10,0,0],[0.25,0.55,0.2])

f =
9.536743164062500e-07
```

Because there are multiple outcomes in this PDF, the plot of the histogram is multidimensional. Just as the binomial distribution histogram (with two outcomes) could be plotted vs one variable, so two can a multinomial histogram with n outcomes, be plotted as a function of n-1 variables, since the last outcome is simply defined by $x_k = n - \sum_{i=1}^{k-1} x_i$.

The histogram for example 4.3 is shown in Figure 4.3.

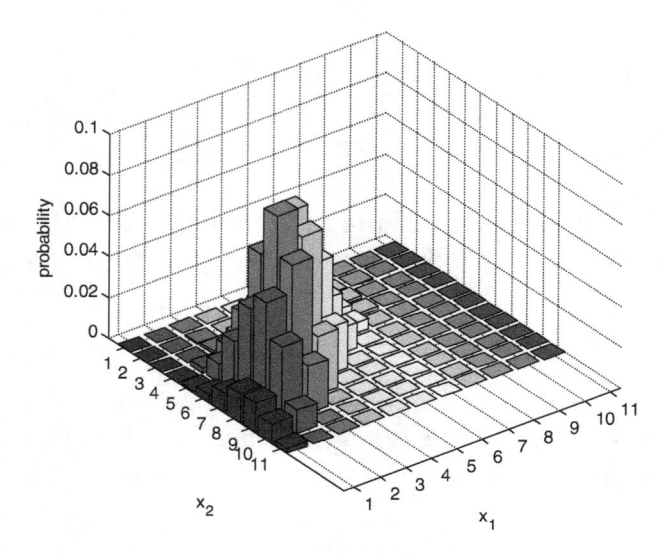

Figure 4.3. The multinomial pdf of the for example 4.3. This PDF has a maximum of 0.0872 at $\underline{x} = [2,6,2]$.

4.6. Hypergeometric Distribution

The hypergeometric distribution applies when

1. A random sample of size n is selected **without replacement** from a sample space containing N total items.
2. k of the N items may be classified as successes and $N-k$ are classified as failures. (Therefore, there are only I outcomes in the experiment.)

The hypergeometric distribution differs from the binomial distribution because the probability of success changes with each trial during the course of the experiment in the hypergeometric case. By contrast, p is constant in the binomial case.

The probability distribution of the hypergeometric random variable X, the number of successes in a random sample of size n selected from a sample space containing a total of N items, in which k of N are will be labeled as a success and $N - k$ will be labeled failure is

$$h(x; N, n, k) = \frac{\binom{k}{x}\binom{N-k}{n-x}}{\binom{N}{n}} \quad \text{for } x = 0,1,2...n \qquad (4.9)$$

The mean of a random variable that follows the hypergeometric distribution $h(x;N,n,k)$ is

$$\mu_x = \frac{nk}{N} \qquad (4.10)$$

and the variance of a random variable that follows the hypergeometric distribution

$$\sigma_x^2 = \left(\frac{N-n}{N-1}\right)\frac{nk}{N}\left(1-\frac{k}{N}\right) \qquad (4.11)$$

Example 4.5: What is the probability of getting dealt 4 of a kind in a hand of five-card-stud poker?
   We can use the hypergeometric distribution on this problem because we are going to select $n=5$ from $N=52$. We don't care about what value of the cards the four of a kind is in so we can just calculate the result for aces and then multiply that probability by 13 since there are 13 values of cards and a four of a kind of any of them are equally likely. Therefore, a success is an ace and a failure is not an ace. $k = 4$ aces. For a four of kind $x = 4$ aces.

$$h(4;52,5,4) = \frac{\binom{4}{4}\binom{52-4}{5-4}}{\binom{52}{5}} = \frac{\binom{4}{4}\binom{48}{1}}{\binom{52}{5}} = \frac{1}{54145}$$

We multiply that by 13 to get the probability of a four-of-a-kind as being: 0.00024. Or, one out of every 4165 hands dealt is probably a four of a kind.

We can also use modern computational tools to evaluate the hypergeometric PDF and CDF. Again, we first show a simple code for the PDF, then demonstrate the use of the intrinsic MATLAB functions.

We can compute the hypergeometric PDF, $P(X = x) = h(x; N, n, k)$, given in equation (4.9) with the following MATLAB code, hypergeo.m:

```
function prob = hypergeo(x,ntot,nsamp,k)
denom = comb(ntot,nsamp);
numerator = comb(k,x)*comb(ntot-k,nsamp-x);
prob = numerator/denom;
```

As an example, if we want to know what is the probability that we draw 3 red marbles in four attempts from a bag containing 5 red marbles and 6 green marbles, then we have $x=3$, $n=4$, $k=5$ and $N=5+6=11$.

```
>> f = hypergeo(3,11,4,5)
```

which returns a value

```
f = 0.181818181818182
```

We can also use the `pdf` function in MATLAB

```
>> f = pdf('Hypergeometric',x,N,k,n)
```

For example, to generate the value of the hypergeometric PDF defined for $x=3$, $n=4$, $k=5$ and $N=11$, one can type the command,

```
>> f = pdf('Hypergeometric',3,11,5,4)
```

which yields the following output

```
f = 0.181818181818182
```

Discrete Probability Distributions - 89

This tells us that the probability of getting 3 red marbles when drawing 4 marbles from a bag initially containing 11 total marbles, 5 of which are red, is about 0.182.

In order to evaluate a cumulative PDF or CDF at a given value of x, one can use the `cdf` function in MATLAB. If we are interested in the probability that $x \leq r$, then the appropriate function is the cumulative distribution function, cdf.

$$F = p(x \leq r) \equiv \sum_{x=0}^{r} h(x; N, n, k)$$

```
>> F = cdf('Hypergeometric',x,N,k,n)
```

For example, to generate the value of the binomial CDF for $x \leq 3$, $n=4$, $k=5$ and $N=11$, one can type the command,

```
>> f = cdf('Hypergeometric',3,11,5,4)
```

which yields the following output

```
f = 0.984848484848485
```

Thus there is a high probability (98.4%) that we draw three of four marbles drawn are red, given that the bag contained only 11 marbles and less than half of them were red.

An example of a histogram of the binomial distribution is shown in Figure 4.4.

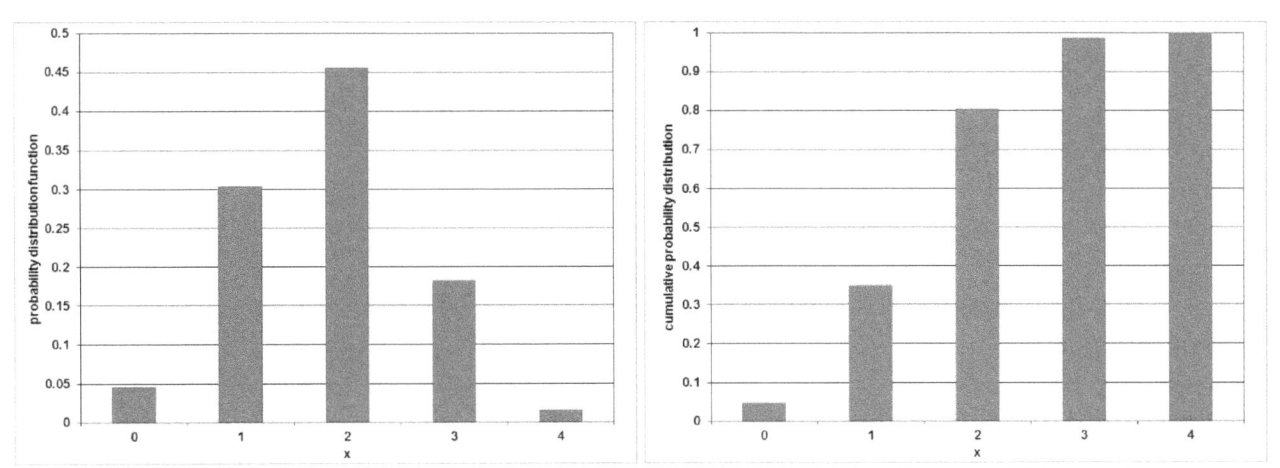

Figure 4.4. The hypergeometric PDF (left) and CDF (right) for a random variable, x, given parameters $n=4$, $k=5$ and $N=11$.

Example 4.6.: A quality control engineer selects a sample of $n=3$ screws from a box containing $N=100$ screws. Of these, 100 screws $k=10$ are defective. What is the probability distribution for $X =$ the number of defective screws that the quality control engineer finds?

First, we see that we can use the hypergeometric distribution because we select n from N, with k of N defined as successes (in this case the detection of a defect). Second, to get the probability distribution we need to find $h(x;N,n,k)$ for all values of x. Since x ranges from 0 to n, we have to solve equation (4.9) for $x = 0, 1, 2,$ and 3.

$$h(x = 0;100,3,10) = \frac{\binom{10}{0}\binom{100-10}{3-0}}{\binom{100}{3}} = \frac{\binom{10}{0}\binom{90}{3}}{\binom{100}{3}} = 0.7265$$

$$h(x = 1;100,3,10) = \frac{\binom{10}{1}\binom{100-10}{3-1}}{\binom{100}{3}} = \frac{\binom{10}{1}\binom{90}{2}}{\binom{100}{3}} = 0.2477$$

$$h(x = 2;100,3,10) = \frac{\binom{10}{2}\binom{100-10}{3-2}}{\binom{100}{3}} = \frac{\binom{10}{2}\binom{90}{1}}{\binom{100}{3}} = 0.0250$$

$$h(x = 3;100,3,10) = \frac{\binom{10}{3}\binom{100-10}{3-3}}{\binom{100}{3}} = \frac{\binom{10}{3}\binom{90}{0}}{\binom{100}{3}} = 0.0007$$

The sum of these probabilities = 0.9999, which is close enough to one, seeing as we only used 4 significant figures.

Binomial Approximation to the hypergeometric distribution

If the population size, N, gets too large in the hypergeometric distribution then we will have problems calculating $N!$. However, if the population gets so large, then whether experiment with or without replacement makes less difference. You can see that if the population was infinitely large, replacement would make no difference at all. For large samples, we can approximate the hypergeometric distribution by the binomial distribution. In this case the hypergeometric parameters and variables:

$$h(x; N, n, k) \approx b(x; n, p = \frac{k}{N})$$

where we see that probability, p, is estimated as the fraction of the N elements that are defined as success.

4.7. Multivariate Hypergeometric Distribution

Just as the binomial distribution can be adjusted to account for multiple random variables (i.e. the multinomial distribution) so too can the hypergeometric distribution account for multiple random variables (multivariate hypergeometric distribution). The multivariate hypergeometric distribution is defined as

$$h_m(\underline{x}; N, n, \underline{a}, k) = \frac{\binom{a_1}{x_1}\binom{a_2}{x_2}\binom{a_3}{x_3}\cdots\binom{a_k}{x_k}}{\binom{N}{n}} \tag{4.12}$$

where x_i is the number of outcomes of the ith result in n trials and a_i is the number of objects of the ith type in the total population of N, and k is the number of types of outcomes. Clearly, the following constraints apply. The total number of each outcome must equal the number of trials, namely $\sum_{i=1}^{k} x_i = n$. The total number of objects must equal the total population, $\sum_{i=1}^{k} a_i = N$.

We can use modern computational tools to evaluate the multivariate hypergeometric distribution. We can write a code to evaluate $P(\{X = x\}) = h_m(\underline{x}; N, n, \underline{a}, k)$, such as the one in multihypergeo.m

```
function prob = multihypergeo(x,ntot,nsamp,a,k)
denom = comb(ntot,nsamp);
numerator = 1.0;
for i = 1:1:k
    numerator = numerator*comb(a(i),x(i));
end
prob = numerator/denom;
```

In the code, multihypergeo.m, x and a are vectors of length k. As an example, if we want to know what is the probability that we draw 3 red marbles and 1 green marble in 4 attempts from a bag containing 5 red marbles and 6 green marbles and 2 blue marbles, then we have $\underline{x} = [3,1,0]$, $N=5+6+2=13$, n=4, $\underline{a} = [5,6,2]$ and $k=3$.

```
>> f = multihypergeo([3,1,0],13,4,[5,6,2],3)
```

which returns a value

```
f = 0.083916083916084
```

MATLAB doesn't have an intrinsic function for the multivariate hypergeometric distribution.

Example 4.7.: A unethical vendor has some defective computer merchandise that he is trying to unload. He has 24 computers. Of these, 12 are ok, 4 others have bad motherboards, 2 others have bad video cards, and 8 others have bad sound cards. If we go into buy 5 computers from this vendor, what is the probability we get 3 good computers, 1 with a bad sound card and 1 with a bad video card?

In this problem, we have $\underline{x} = [3,0,1,1]$, $N=24$, $n=5$, $\underline{a} = [12,4,2,8]$ and $k=4$.

$$P(\{X\} = [3,0,1,1]) = h_m([3,0,1,1];24,5,[12,4,2,8],k) = \frac{\binom{12}{3}\binom{4}{0}\binom{2}{1}\binom{8}{1}}{\binom{24}{5}} = 0.08282$$

So there is about an 8.2% chance that when we purchase five computers from this vendor that we get 3 good computers, 1 with a bad sound card and 1 with a bad video card.

4.8. Negative Binomial Distribution

Like the binomial distribution, the negative binomial distribution applies to a Bernoulli process, but it asks a different question. As a reminder the Bernoulli process satisfies the following three criteria:

1. An experiment that consists of x repeated, **independent** trials.
2. Each trial can have one of two outcomes, success or failure.
3. The probability of success, p, is the same for each trial.

The negative binomial distribution applies when a fourth criteria is added.

4. The trials are continued until we achieve the kth success.

This is the Bernoulli process, except that in the binomial distribution, we fixed *n* trials and allowed *x*, the number of successes to be the random variable. In the negative binomial distribution, we fix k successes and allow the number of trials, now labelled x, to be the random variable.

The probability distribution of the negative binomial random variable X, the number of trials needed to obtain k successes is

$$b^*(x;k,p) = \binom{x-1}{k-1} p^k q^{x-k} \quad \text{for } x = k, k+1, k+2... \tag{4.13}$$

Remember the probability of failure is one less the probability of success, $q = 1-p$.

We can also use modern computational tools to evaluate the negative binomial PDF and CDF. Again, we first show a simple code for the PDF, then demonstrate the use of the intrinsic MATLAB functions.

We can compute the negative binomial PDF, $P(X = x) = b^*(x;k,p)$, given in equation (4.13) with the following MATLAB code, negbinomial.m:

```
function prob = negbinomial(x,k,p)
prob = comb(x-1,k-1)*p^k*(1-p)^(x-k);
```

As an example, if we want to know what is the probability that we draw flip our second head on our fourth trial, where p=0.5, we have x=4 and k=2

```
>> f = negbinomial(4,2,0.5)
```

which returns a value

```
f = 0.187500000000000
```

We can also use the `pdf` function in MATLAB

```
>> f = pdf('Negative Binomial',x-k,k,p)
```

This formulation with the random variable being x-k, rather than x is due to alternate formulations of the negative binomial distribution, in which the function assumes the random variable is the number of failures, not the number of trials. The number of failures is the number of trials less the number of successes. For example, to generate the value of the negative binomial PDF defined for x=4, k=2 and p=0.5, one can type the command,

```
>> f = pdf('Negative Binomial',4-2,2,0.5)
```

which yields the following output

```
f = 0.187500000000000
```

In order to evaluate a cumulative PDF or CDF at a given value of x, one can use the `cdf` function in MATLAB. If we are interested in the probability that $x \leq r$, then the appropriate function is the cumulative distribution function.

$$F = p(x \leq r) \equiv \sum_{x=k}^{r} b^*(x; k, p)$$

```
>> F = cdf('Negative Binomial',x-k,k,p)
```

For example, to generate the value of the negative binomial PDF defined for $x \leq 4$, $k=2$ and $p=0.5$, one can type the command,

```
>> F = cdf('Negative Binomial',4-2,2,0.5)
```

which yields the following output

```
F = 0.687500000000000
```

Thus the probability that we achieve our second success in less than four trials is 68.75%.

An example of a histogram of the negative binomial distribution is shown in Figure 4.5.

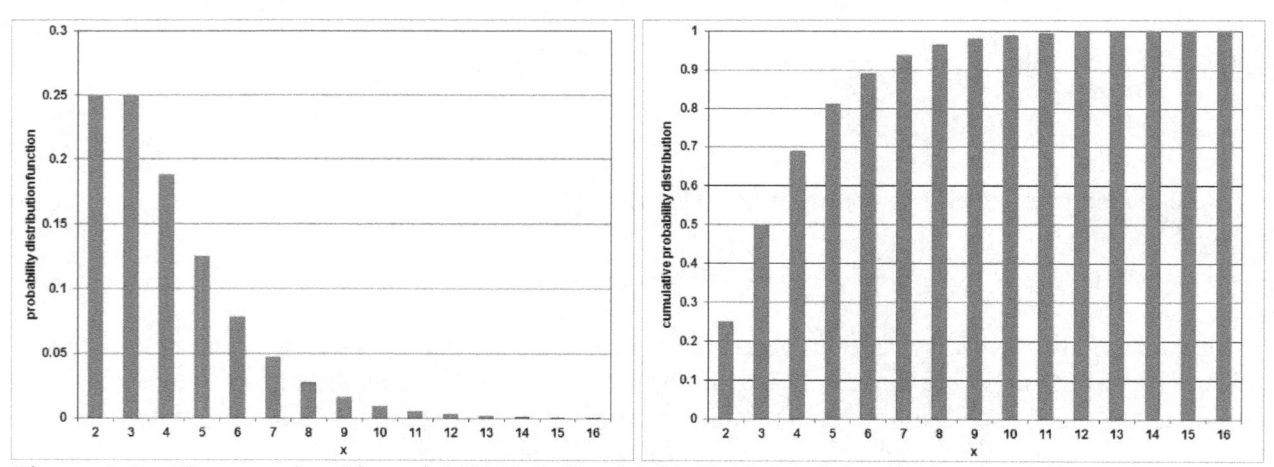

Figure 4.5. The negative binomial PDF (left) and CDF (right) for a random variable, x, given parameters $k=2$ and $p=0.5$.

Example 4.8.: What is the probability when flipping four coins of getting all heads or all tails for the second time on the fifth toss?

Here we can use the negative binomial because we know that we want the $k=2$ success, for an independent trial with $p = 1/8$ for the specific case where $x = 4$.

$$b^*(5;2,1/8) = \binom{5-1}{2-1}\left(\frac{1}{8}\right)^2\left(\frac{7}{8}\right)^3 = 0.04187$$

```
>> f = pdf('Negative Binomial',5-2,2,1/8)
```

which yields the following output

```
f = 0.041870117187500
```

So there is 4.2% chance when flipping four coins of getting all heads or all tails for the second time on the fifth trial.

4.9. Geometric Distribution

The geometric distribution is a subset of the negative binomial distribution when $k=1$. That is, the geometric distribution gives the probability that the first success occurs on the random variable X, the number of the trial. This corresponds to a system where the Bernoulli process stops after the first success. If success is defined as a "failure" of the system, a crash of a code, or an explosion, or some fault in the process, which causes the system to stop functioning, then in this case, we are only interested in one such success, so $k=1$ and we have the geometric distribution.

The probability distribution of the geometric random variable X, the number of trials needed to obtain the first success is

$$g(x;p) = pq^{x-1} \quad \text{for } x = 1,2,3... \tag{4.14}$$

The mean of a random variable following the geometric distribution is

$$\mu_x = \frac{1}{p} \tag{4.15}$$

and the variance is

$$\sigma_x^2 = \frac{1-p}{p^2} \tag{4.16}$$

Remember the probability of failure is one less the probability of success, $q = 1-p$.

We can also use modern computational tools to evaluate the geometric PDF and CDF. Again, we first show a simple code for the PDF, then demonstrate the use of the intrinsic MATLAB functions.

We can compute the geometric PDF, $P(X = x) = g(x; p)$, given in equation (4.14) with the following MATLAB code, geo.m:

```
function prob = geo(x,p)
prob = p*(1-p)^(x-1);
```

As an example, if we want to know what is the probability that we draw flip our first head on our fourth trial, where $p=0.5$, we have $x=4$ and $k=1$

```
>> f = geo(4,0.5)
```

which returns a value

```
f = 0.062500000000000
```

We can also use the `pdf` function in MATLAB

```
>> f = pdf('Geometric',x-1,p)
```

This formulation with the random variable being $x-1$, rather than x is due to alternate formulations of the geometric distribution, in which the function assumes the random variable is the number of failures, not the number of trials. The number of failures is the number of trials less the number of successes. For example, to generate the value of the negative binomial PDF defined for $x=4$, $k=2$ and $p=0.5$, one can type the command,

```
>> f = pdf('Geometric',4-1,0.5)
```

which yields the following output

```
f = 0.062500000000000
```

In order to evaluate a cumulative PDF or CDF at a given value of x, one can use the `cdf` function in MATLAB. If we are interested in the probability that $x \leq r$, then the appropriate function is the cumulative distribution function.

$$F = p(x \leq r) \equiv \sum_{x=1}^{r} g(x; p)$$

```
>> F = cdf('Geometric',x-1,p)
```

For example, to generate the value of the negative binomial PDF defined for $x \leq 4$ and $p=0.5$, one can type the command,

```
>> F = cdf('Geometric',4-1,0.5)
```

which yields the following output

```
F = 0.937500000000000
```

Thus the probability that we achieve our first success in less than four trials is 93.75%.

An example of a histogram of the negative binomial distribution is shown in Figure 4.6.

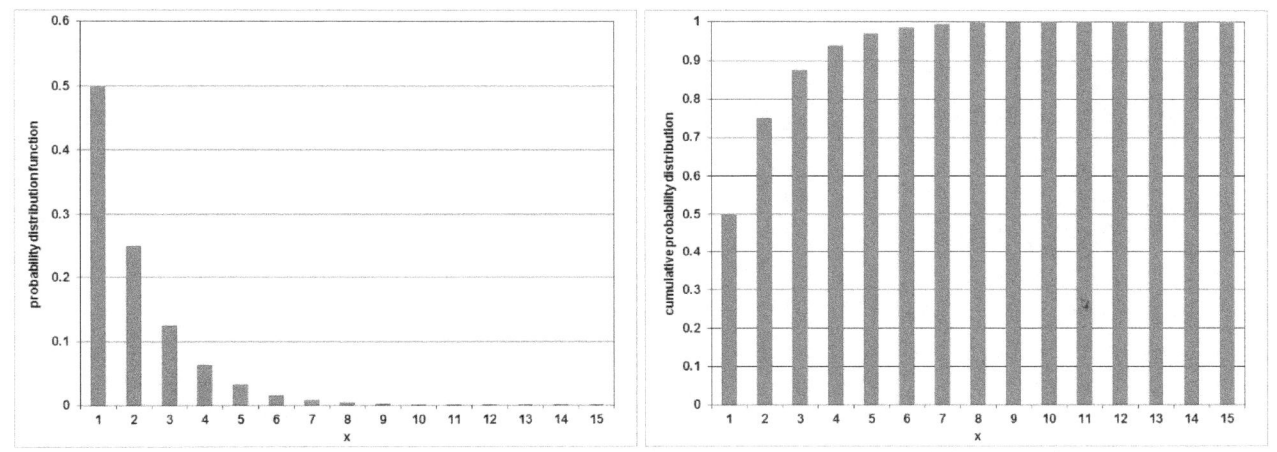

Figure 4.6. The geometric PDF (left) and CDF (right) for a random variable, x, given parameter $p=0.5$.

Example 4.9.: A recalcitrant child is told he cannot leave the dinner table until she eats one pea. Each time the child brings that ominous pea close to her mouth, there is a 90% chance her will crumbles and the spoon shakes, and the pea falls to the floor, where it is gobbled up by the family dog, forcing the child to try again. What is the probability that the child eats the pea on the first through fiftieth try?

For $x = 1$, $g(x = 1; p = 0.1) = (0.1)(0.9)^{1-1} = 0.1$. Similar calculations yield the distribution plotted in Figure 4.7. The cumulative distribution in Figure 4.7 gives the probability that the pea has been eaten by the attempt x.

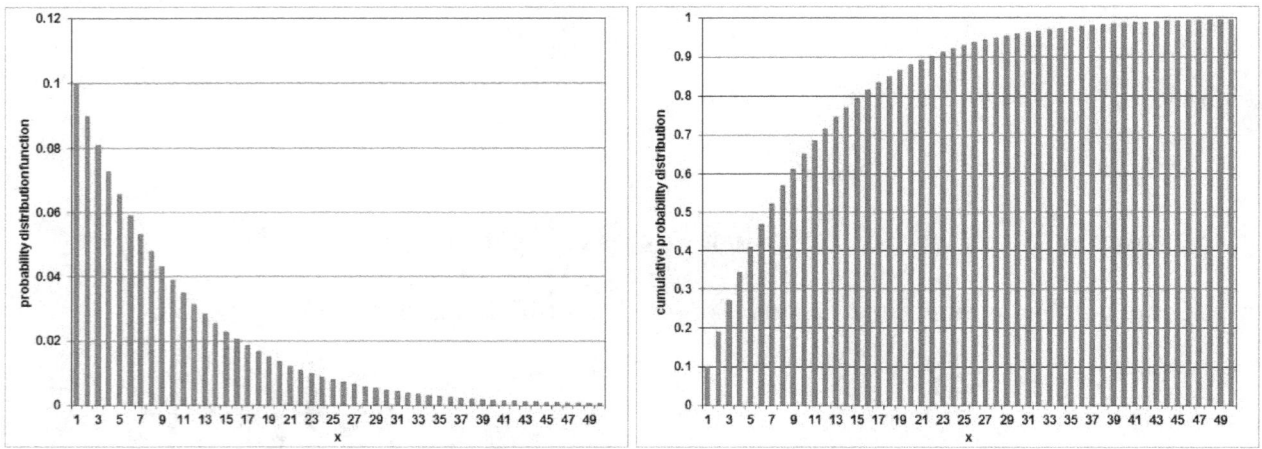

Figure 4.7. The geometric PDF (left) and CDF (right) for a random variable, *x*, given parameter *p*=0.1.

4.10. Poisson Distribution

When we looked at binomial, negative binomial, hypergeometric, geometric distributions, we had two outcomes success and failure. In the Poisson distribution, the random variable *X* is a number. In fact, it is the number of outcomes (no longer classified as a success or failure) during a given interval (of time or space). For example, the random variable *X* could be the number of baseball games postponed due to rain in a baseball season, or the number of bacteria in a petri dish.

The Poisson process is a collection of Poisson experiments, with the properties

1. The number of outcomes in one interval is independent of the number that occurs in any disjoint interval.
2. The probability that a single outcome will occur during a very short interval is proportional to the length of the interval and does not depend on the number of outcomes outside the interval.
3. The probability that more than one outcome will occur in an infinitesimally small interval is negligible.

In the Poisson distribution, *t* is the size of the interval, λ is the rate of the occurrence of the outcome, and *X* is the number of outcomes occurring in interval *t*. The probability distribution of the Poisson random variable *X* is

$$p(x; \lambda t) = \frac{e^{-\lambda t}(\lambda t)^x}{x!} \text{ for } x = 0,1,2... \tag{4.17}$$

The cumulative probability distribution, that is the probability for getting anywhere between 0 and *r* outcomes, inclusive is

$$P(r;\lambda t) = \sum_{x=0}^{r} p(x;\lambda t) \qquad (4.18)$$

The mean and the variance of the Poisson distribution are

$$\mu_x = \sigma_x^2 = \lambda t \qquad (4.19)$$

The Poisson distribution is the asymptotical form of the binomial distribution when n, the number of trials, goes to infinity, p, the probability of a success goes to zero, and the mean (np) remains constant. We do not prove this relation here.

We can also use modern computational tools to evaluate the geometric PDF and CDF. Again, we first show a simple code for the PDF, then demonstrate the use of the intrinsic MATLAB functions.

We can compute the Poisson PDF, $P(X = x) = p(x;\lambda t)$, given in equation (4.17) with the following MATLAB code, poisson.m:

```
function f = poisson(x,p)
f= exp(-p)*p^x/factorial(x);
```

As an example, if we want to know what is the probability that we observe $x = 2$ outcomes in time span of $t = 3$ days given that the average rate of outcomes per day is given by $\lambda = 0.2$, we have

```
>> f = poisson(2,0.2*3)
```

which returns a value

```
f = 0.098786094496925
```

We can also use the pdf function in MATLAB

```
>> f = pdf('Poisson',x,p)
```

For example, to generate the value of the Poisson PDF defined for $x=2$, $t=3$ and $\lambda=0.2$, one can type the command,

```
>> f = pdf('Poisson',2,0.2*3)
```

which yields the following output

```
f = 0.098786094496925
```

In order to evaluate a cumulative PDF or CDF at a given value of x, one can use the `cdf` function in MATLAB. If we are interested in the probability that $x \leq r$, then the appropriate function is the cumulative distribution function.

$$F = p(x \leq r) \equiv \sum_{x=0}^{r} p(x; \lambda t)$$

```
>> F = cdf('Poisson',x,p)
```

For example, Poisson PDF defined for $x \leq 2$, $t=3$ and $\lambda=0.2$, one can type the command,

```
>> F = cdf('Poisson',2,0.2*3)
```

which yields the following output

```
F = 0.976884712247367
```

Thus the probability that we observe two or less outcomes in three days is about 97.7%.

An example of a histogram of the Poisson distribution is shown in Figure 4.8.

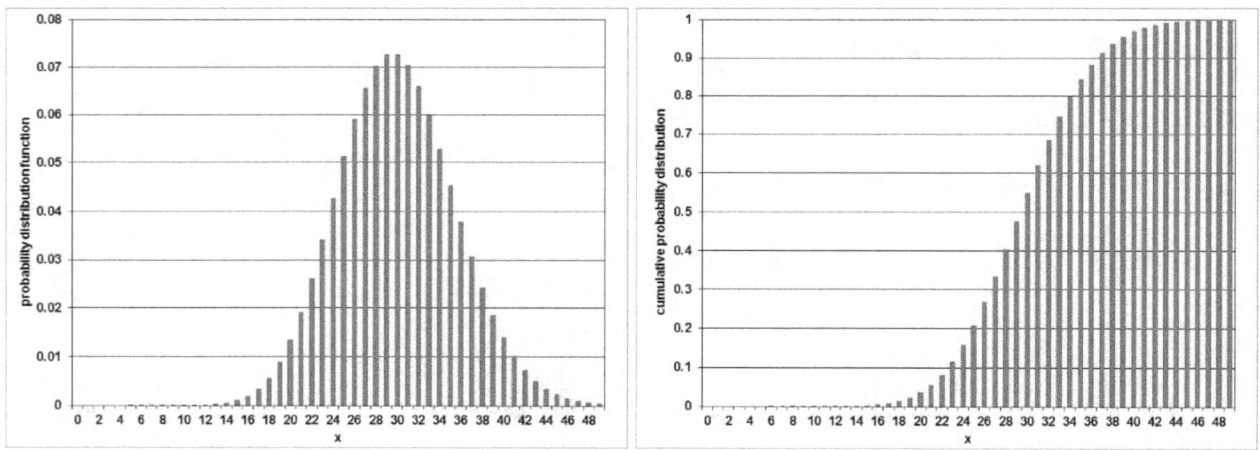

Figure 4.8. The geometric PDF (left) and CDF (right) for a random variable, x, given parameters $t=3$ and $\lambda=10$.

Example 4.10.: Historical quality control studies at a plant indicate that there is a defect rate of 1 in a thousand products. What is the probability that in 10000 products there are exactly 5 defects? Less than or equal to 5 defects?

Using equation (4.17), we have:

$$p(x=5; \lambda t) = \frac{e^{-\lambda t}(\lambda t)^x}{x!} = \frac{e^{-10}(10)^5}{5!} = 0.0378$$

Alternatively, using the MATLAB pdf function, we have

```
>> f = pdf('Poisson',5,10)
```

which yields the following output

```
f =  0.037833274802071
```

The cumulative pdf yields the probability that x is less than or equal to 5.

```
>> F = cdf('Poisson',5,10)
```

which yields the following output

```
f =  0.067085962879032
```

So there is about a 6.7% chance of observing 5 or fewer defects in 10,000 products.

4.11. Subroutines

Code 4.1. Binomial probability distribution (binomial.m)
This code, binomial.m, provides the binomial probability distribution. Note that it calls comb.m from Code 1.3.

```
function f = binomial(x,n,p)
f = comb(n,x)*p^x*(1-p)^(n-x);
```

Code 4.2. Cumulative Binomial probability distribution (binocumu.m)
This code, binocumu.m, provides the cumulative binomial probability distribution. Note that it calls biniomial.m (Code 4.1), which calls comb.m (Code 1.3).

```
function f = binocumu(r,n,p)
f = 0.0
for x = 0:1:r
   f = f + binomial(x,n,p);
end
```

Code 4.3. Arbitrary Ranges of Binomial probability distribution (binoprob.m)

This code, binoprob.m, provides the binomial probability distribution over arbitrary ranges of x. Note that it calls biniomial.m (Code 4.1), which calls comb.m (Code 1.3).

```
function f = binoprob(a,c,n,p)
f = 0.0
for x = a:1:c
   f = f + binomial(x,n,p);
end
```

Code 4.4. Multinomial probability distribution (multinomial.m)

This code, multinomial.m, provides the multinomial probability distribution.

```
function prob = multinomial(x,n,p,k)
prob = factorial(n);
for i = 1:1:k
   prob = prob/factorial(x(i))*p(i)^x(i);
end
```

Code 4.5. Hypergeometric probability distribution (hypergeo.m)

This code, hypergeo.m, provides the hypergeometric probability distribution. Note that it calls comb.m from Code 1.3.

```
function prob = hypergeo(x,ntot,nsamp,k)
denom = comb(ntot,nsamp);
numerator = comb(k,x)*comb(ntot-k,nsamp-x);
prob = numerator/denom;
```

Code 4.6. Multivariate Hypergeometric probability distribution (multihypergeo.m)

This code, multihypergeo.m, provides the multivariate hypergeometric probability distribution. Note that it calls comb.m from Code 1.3.

```
function prob = multihypergeo(x,ntot,nsamp,a,k)
denom = comb(ntot,nsamp);
numerator = 1.0;
for i = 1:1:k
   numerator = numerator*comb(a(i),x(i));
end
prob = numerator/denom;
```

Discrete Probability Distributions - 103

Code 4.7. Negative Binomial probability distribution (negbinomial.m)

This code, negbinomial.m, provides the negative binomial probability distribution. Note that it calls comb.m from Code 1.3.

```
function prob = negbinomial(x,k,p)
prob = comb(x-1,k-1)*p^k*(1-p)^(x-k);
```

Code 4.8. Geometric probability distribution (geo.m)

This code, geo.m, provides the geometric probability distribution.

```
function prob = geo(x,p)
prob = p*(1-p)^(x-1);
```

Code 4.9. Arbitrary Ranges of the Geometric probability distribution (geoprob.m)

This code, geoprob.m, provides the geometric probability distribution over an arbitrary range of the random variable, in order to evaluate $P(a \leq X \leq c) = \sum_{i=a}^{c} g(x;p)$.

```
function f = geoprob(a,c,p)
f = 0.0;
for x = a:1:c
f = f + geo(x,p);
end
```

Code 4.10. Poisson probability distribution (poisson.m)

This code, geo.m, provides the geometric probability distribution.

```
function f = poisson(x,p)
f= exp(-p)*p^x/factorial(x);
```

Code 4.11. Arbitrary Ranges of the Poisson probability distribution (poisprob.m)

This code, poisprob.m, provides the Poisson probability distribution over an arbitrary range of the random variable, in order to evaluate $P(a \leq X \leq c) = \sum_{x=a}^{c} p(x;p)$.

```
function f = poisprob(a,c,p)
f = 0.0;
for x = a:1:c
f = f + poisson(x,p);
end
```

4.12. Problems

Problem 4.1.
You have a computer code that generates random integers in the range 20 to 40, inclusive.
(a) If the random variable, x, is the value of the random integer, what PDF describes the distribution of x?
(b) What is the probability that 25<x<35?
(c) What is the mean value of the random integer?
(d) What is the variance of the integer?

Problem 4.2.
According to Chemical Engineering Progress (Nov. 1990) approximately 30% of all pipework failures in chemical plants are caused by operator error.
(a) If x is a random variable that describes the number of pipework failures caused by operator error, what PDF will describe x?
(b) What is the probability that out of the next 20 pipework failures, at least 10 are due to operator error?
(c) What is the probability that no more than 4 out of 20 such failures are due to operator error?
(d) What is the probability that out of the next 20 pipework failures, exactly 5 are due to operator error?

Problem 4.3.
According to genetics theory, a certain cross of guinea pigs will result in red, black, and white offspring in the ratio 8:4:8.
(a) If {x} is a set of random variables that describes the number of red, black, and white offspring, what PDF describes {x}?
(b) Find the probability that among 8 offspring 5 will be red, 2 black, and 1 white.

Problem 4.4.
An urn contains 3 green balls, 2 blue balls, and 4 red balls. A random sample of 5 balls is selected.
(a) If {x} is a set of random variables that describes the number of balls of each color drawn, what PDF describes {x}?
(b) Find the probability that in the sample of 5 balls, both blue balls and at least 1 red ball are selected.

Problem 4.5.
Population studies of biology and the environment often tag and release subjects in order to estimate size and degree of certain features in the population. 10 animals of a certain population thought to be near extinction are caught, tagged and released in a certain region. After a period of time, a random sample of 15 of the types of animals are caught. There are only 25 of the animals in the region.

(a) If x is a random variable that describes the number of animals caught both times, what PDF describes x?

(b) What is the probability that five of the animals caught in the second batch had been caught in the first batch?

Problem 4.6.

Among 150 IRS employees in a large city, only 30 are women. (a) If 10 of the employees are chosen at random to provide free tax assistance to residents of the city, use the binomial approximation to the hypergeometric PDF to find the probability that at least three women were selected.

Problem 4.7.

A scientist inoculates several mice, one at a time, with a disease germ until he finds 2 that have contracted the disease. If the probability of contracting the disease is 1/6.

(a) If x is a random variable that describes the number of mice which must be inoculated, what PDF describes x?

(b) What is the probability that 8 mice are required?

(c) What is the probability that between 8 and 10 mice, inclusive, are required?

Problem 4.8.

The acceptance scheme for purchasing lots containing a large number of batteries is to test no more than 75 randomly selected batteries and to reject a lot if a single battery fails. Suppose the probability of a failure is 0.001.

(a) What is the probability that a lot is accepted?

(b) What is the probability that a lot is rejected on the 20th attempt?

(c) What is the probability that it is rejected in 10 or less trials?

Problem 4.9.

On average a certain intersection results in 3 traffic accidents per month. What is the probability that for any given month at this intersection

(a) exactly 5 accidents occur?

(b) less than 3 accidents occur?

(c) at least 2 accidents occur?

Chapter 5. Continuous Probability Distributions

5.1. Introduction

In this chapter, we introduce some of the common probability density functions (PDFs) for continuous sample spaces. The goal of this section is to become familiar with these PDFs and to be able to easily evaluate quantities associated with the PDFs. When given a word problem, we will also learn how to identify which PDF is appropriate.

This chapter is organized with each section describing a different PDF, including continuous, normal, Student's t, gamma, exponential, chi-squared and F distributions. To be clear this is simply a small set of continuous PDFs with common application in the sciences and in statistics. A final section discusses functions of random variables.

5.2. Continuous Uniform Distribution

The continuous uniform distribution describes a sample space where every point within the space is equally (uniformly) likely.

The density function of the continuous uniform random variable X on the interval [A,B] is

$$f(x;A,B) = \begin{cases} \dfrac{1}{B-A} & \text{for } A \leq x \leq B \\ 0 & \text{otherwise} \end{cases} \quad (5.1)$$

For any of the continuous PDF, the population mean of any function of a random variable, $g(x)$, is given by the definition first provided in Chapter 3.

$$\mu_{g(x)} \equiv \int_{-\infty}^{\infty} g(x) f(x) dx \quad (3.\text{X})$$

The application of this definition to the mean of x for the continuous uniform PDF leads to

$$\mu_x = \int_A^B x \frac{1}{B-A} dx = \frac{1}{B-A} \int_A^B x\, dx = \frac{1}{B-A}\left[\frac{x^2}{2}\right]_A^B = \frac{1}{B-A}\frac{B^2-A^2}{2} \tag{5.2}$$

Further simplification leads to an intuitive formula for the mean of x for the continuous uniform PDF

$$\mu_x = \frac{B+A}{2} \tag{5.3}$$

For any of the continuous PDF, the population variance of any function of a random variable, $g(x)$, is given by the definition first provided in Chapter 3.

$$\sigma^2_{g(x)} \equiv \int_{-\infty}^{\infty} [g(x) - \mu_{g(x)}]^2 f(x)\, dx \tag{3.Y}$$

The application of this definition to the variance of x for the continuous uniform PDF leads to

$$\sigma^2_x = \int_A^B (x-\mu_x)^2 \frac{1}{B-A} dx = \frac{1}{B-A}\left[\frac{(x-\mu_x)^3}{3}\right]_A^B = \frac{1}{B-A}\left[\frac{(B-\mu_x)^3}{3} - \frac{(A-\mu_x)^3}{3}\right] \tag{5.4}$$

Further simplification leads to a formula for the variance of x for the continuous uniform PDF

$$\sigma^2_x = \frac{(B-A)^2}{12} \tag{5.5}$$

The Continuous Uniform Probability Distribution in MATLAB

There are a variety of useful tools for dealing with continuous distributions in MATLAB. A few of them are reviewed here with respect to their application to the continuous uniform distribution.

In order to generate a random number from the continuous uniform PDF defined from A to B, one can use the `random` function in MATLAB,

```
>> x = random('Uniform',A,B)
```

For example, to generate a random number from the continuous uniform PDF defined from -1 to 1, one can type the command,

```
>> x = random('Uniform',-1,1)
```

which yields an output such as

```
x =   -0.804919190001181
```

In order to evaluate a PDF at a given value of x, one can use the `pdf` function in MATLAB

```
>> f = pdf('Uniform',x,A,B)
```

For example, to generate the value of the PDF for the continuous uniform PDF defined from -1 to 1 at $x = 0.3$, one can type the command,

```
>> f = pdf('Uniform',0.3,-1,1)
```

which yields the following output

```
f =    0.500000000000000
```

In order to evaluate a cumulative PDF at a given value of x, one can use the `cdf` function in MATLAB. If we are interested in the probability that $x \leq x_{hi}$, then the appropriate function is the cumulative distribution function.

$$F = p(x \leq x_{hi}) \equiv \int_{-\infty}^{x_{hi}} f(x)dx$$

In MATLAB, we can directly evaluate the cumulative distribution function for a number of common PDFs, including all of the continuous PDFs studies in this course.

```
>> F = cdf('Uniform',xhi,A,B)
```

For example, to generate the value of the cumulative PDF for the continuous uniform PDF defined from -1 to 1 at $x = 0.3$, one can type the command,

```
>> F = cdf('Uniform',0.3,-1,1)
```

which yields the following output

```
F =    0.650000000000000
```

In other words,

Continuous Probability Distributions - 109

$$F = p(x \leq 0.3) \equiv \int_{-\infty}^{0.3} f(x;)dx = 0.65$$

Above we were given x_{hi} and asked to find the probability, p. Frequently, we are asked the inverse problem, where we are given, p, and asked to find the value of x_{hi}. In order to determine the value of a random variable x at which the cumulative PDF of the continuous uniform distribution has a value of F, one can use the `icdf` (the inverse cdf) function in MATLAB

```
>> xhi = icdf('Uniform',F,A,B)
```

For example, to determine the value of a random variable x at which the cumulative uniform PDF, defined from -1 to 1, has a value of 0.6, one can type the command,

```
>> xhi = icdf('Uniform',0.6,-1,1)
```

which yields the following output

```
xhi =     0.200000000000000
```

In other words,

$$F = 0.6 = p(x \leq 0.2) \equiv \int_{-\infty}^{0.2} f(x)dx$$

Example 5.1: The default random number generators for most programming languages provides a uniform distribution on the interval [0,1]. What is the probability that a MATLAB random number generator yields (a) X=0.25 (b) X<0.25, (c) x≤0.25, (d) X>0.25, (e) 0.1<X<0.25?

(a) The probability of getting exactly any number in a continuous sample space is zero.

(b) Here we use the definition of a continuous PDF.

$$P(X < 0.25) = P(0 < X < 0.25) = \int_0^{0.25} f(x)dx = \int_0^{0.25} \frac{1}{1-0}dx = 0.25$$

Alternatively, we can solve this problem in MATLAB.

```
>> F = cdf('Uniform',0.25,0,1)

F =     0.250000000000000
```

(c) The answer for (c) is the same as the answer for (b).

$P(X \leq 0.25) = P(X < 0.25) + P(X = 0.25) = P(X < 0.25) = 0.25$

(d) The continuous uniform PDF is normalized. So, the answer to (d) is given by

$P(X > 0.25) = 1 - P(X < 0.25) = 1 - 0.25 = 0.75$

(e)

$$P(0.1 < X < 0.25) = \int_{0.1}^{0.25} f(x)dx = \int_{0.1}^{0.25} \frac{1}{1-0} dx = 0.15$$

Alternatively, we can solve this problem in MATLAB.

```
>> Fhi = cdf('Uniform',0.25,0,1);
>> Flo = cdf('Uniform',0.1,0,1);
>> p = Fhi-Flo

p =    0.150000000000000
```

5.3. Normal Distribution

The normal distribution is one of the most important continuous PDF in statistics and in science. It is also known as the Gaussian distribution and many other application-specific names. For example, the Maxwell-Boltzmann distribution is the name given to the normal distribution when it is applied to the distribution of atomic velocities in a material. The shape of the normal distribution has given rise to the nickname "the bell curve".

The probability density function of the normal random variable, x, with mean μ and variance σ^2, is

$$f(x;\mu,\sigma) = \frac{1}{\sqrt{2\pi}\sigma} e^{-\frac{1}{2}\left(\frac{x-\mu}{\sigma}\right)^2} \tag{5.6}$$

You will notice immediately that this distribution is different than the PDF's we've studied previously. The normal PDF is defined by two parameters, the mean and variance, μ and σ^2. In our previous cases, the PDF defined μ and σ^2. The effect of the mean on the normal PDF is shown in Figure 5.1 Clearly the normal distribution is centered around the mean.

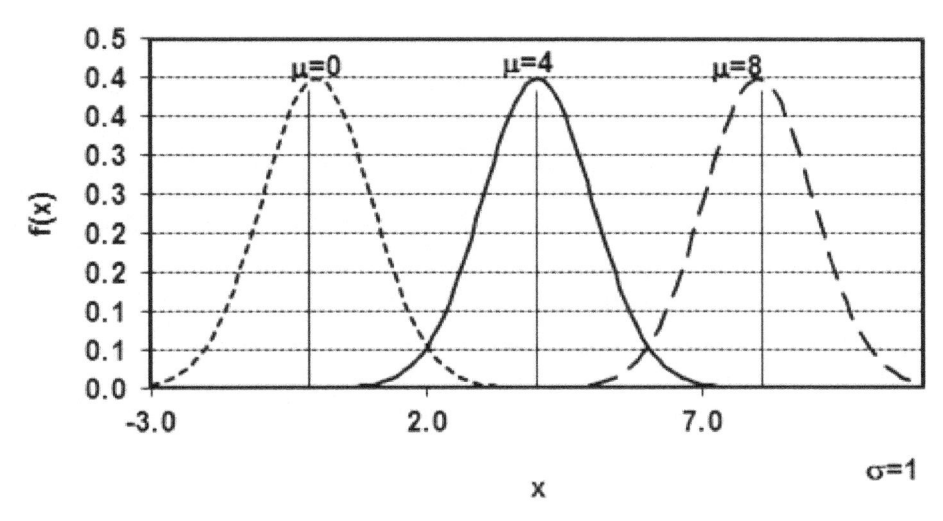

Figure 5. The effect of mean on the normal distribution with a variance of one.

The effect of the variance on the normal PDF is shown in Figure 5.2. Clearly, the normal distribution grows broader as the variance increases.

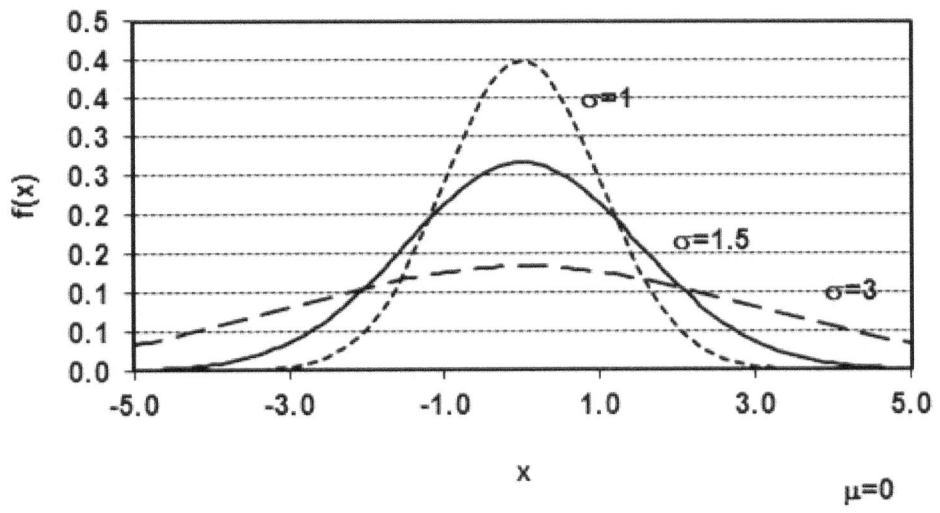

Figure 5.2. The effect of variance on the normal distribution with a mean of zero.

Some characteristics of the Normal PDF include
- The mode, (the most probable value occurs at the mean).
- The distribution is symmetric about the mean.
- The curve has inflection point at $x = \mu \pm \sigma$. (Remember: inflection points are where the second derivative is zero, where the curve changes from concave up to concave down.
- The normal curve approaches the x-axis as we move from the mean.
- The total area under the curve and above the x-axis is one (as it is for all PDF's.)

- The probability $P(a < X < b) = \int_a^b f(x)dx$ is the area under the normal curve between a and b.
  - The Normal PDF with $\mu = 0$ and $\sigma = 1$ is called the **Standard Normal PDF**.
  - Any Normal PDF, $f(x;\mu,\sigma)$, can be converted to a standard normal PDF, $f(z;0,1)$, with the change of variable $z = \dfrac{x-\mu}{\sigma}$.
  - The integral of the normal PDF in equation (5.6) cannot be analytically evaluated.
  - The normal PDF $f(x;\mu,\sigma)$ is equal to the binomial PDF $b(x;n,p) = \binom{n}{x}p^x q^{n-x}$ with $\mu = np$ and $\sigma = npq$ in the limit when $n \to \infty$.

Since the PDF cannot be analytically integrated, tables of integrals for this commonly used distribution are widely available. Typically, the tables are given for the cumulative PDF,

$$P(x \le a) = F(a) = \int_{-\infty}^{a} f(x;\mu,\sigma)dx = \int_{-\infty}^{a} \frac{1}{\sqrt{2\pi}\sigma} e^{-\frac{1}{2}\left(\frac{x-\mu}{\sigma}\right)^2} dx \qquad (5.7)$$

When the pioneers were crossing the plains in their covered wagons and they wanted to evaluate probabilities from the normal distribution, they used tables of the cumulative normal PDF, such as those provided in the back of the Walpole, Myers, Myers and Ye textbook. These tables are also available online. For example wikipedia has a table of cumulative standard normal PDFs at

http://en.wikipedia.org/wiki/Standard_normal_table

These tables provide values of the cumulative standard normal PDF, $f(z;0,1)$, for discretized values of the random variable, z.

The Normal Distribution in MATLAB

Alternatively, one can use tools in MATLAB to evaluate problems concerned with the normal distribution. In order to generate a random number from the normal PDF defined by mean, m, and standard deviation, s, one can use the `random` function in MATLAB,

```
>> x = random('normal',m,s)
```

For example, to generate a random number from the normal PDF defined by mean and standard deviation, 0 and 1, one can type the command,

Continuous Probability Distributions - 113

```
>> x = random('normal',0,1)
```

which yields an output such as

```
x = -2.258846861003648
```

In order to evaluate a PDF at a given value of x, one can use the `pdf` function in MATLAB

```
>> f = pdf('normal',x,m,s)
```

For example, to generate the value of the PDF for the normal PDF defined by mean and standard deviation, 0 and 1 at x = 0.3, one can type the command,

```
>> f = pdf('normal',0.3,0,1)
```

which yields the following output

```
f = 0.381387815460524
```

In order to evaluate a cumulative PDF at a given value of x, one can use the `cdf` function in MATLAB

```
>> F = cdf('normal',xhi,m,s)
```

For example, to generate the value of the cumulative PDF for the normal PDF defined by mean and standard deviation, 0 and 1 at x = 0.3, one can type the command,

```
>> F = cdf('normal',0.3,0,1)
```

which yields the following output

```
F = 0.617911422188953
```

In other words,

$$F = p(x \leq 0.3) \equiv \int_{-\infty}^{0.3} f(x)dx = 0.61791$$

In order to determine the value of a random variable x at which the cumulative PDF of the normal distribution has a value of F, one can use the `icdf` (the inverse cdf) function in MATLAB

```
>> xhi = icdf('normal',F,m,s)
```

For example, to determine the value of the random variable x at which the cumulative normal PDF with mean and standard deviation, 0 and 1, has a value of 0.6, one can type the command,

```
>> xhi = icdf('normal',0.6,0,1)
```

which yields the following output

```
xhi =    0.253347103135800
```

In other words,

$$F = 0.6 = p(x \leq 0.25335) = \int_{-\infty}^{0.25335} f(x)dx$$

Example 5.2: A type of battery lasts on the average 3.0 years with a standard deviation of 0.5 years. What is the probability that the battery will last (a) less than 2.3 years? (b) between 2.5 and 3.5 years? (c) more than four years?

(a) If we solve this problem manually, the first step is to convert the non-standard normal PDF to the standard normal PDF via the transformation

$$z_{2.3} = \frac{x-\mu}{\sigma} = \frac{2.3-3.0}{.5} = -1.4$$

One can then go visit a standard normal table and look up the value corresponding to $z = -1.4$.

$$P(x \leq 2.3) = P(z \leq -1.4) = 0.0808$$

Alternatively, one can solve the problem in MATLAB.

```
>> F = cdf('normal',2.3,3,0.5)

F =    0.080756659233771
```

(b) Manually, we transform both limits.

$$z_{2.5} = \frac{x-\mu}{\sigma} = \frac{2.5-3.0}{.5} = -1.0, \quad z_{3.5} = \frac{x-\mu}{\sigma} = \frac{3.5-3.0}{.5} = 1.0$$

We use the cumulative table twice and take the difference.

Continuous Probability Distributions - 115

$$P(2.5 \leq X \leq 3.5) = P(-1.0 \leq Z \leq 1.0) = P(Z \leq 1.0) - P(Z \leq -1.0)$$
$$= 0.8413 - 0.1587 = 0.6826$$

Alternatively, one can solve the problem in MATLAB.

```
>> Fhi = cdf('normal',3.5,3,0.5);
>> Flo = cdf('normal',2.5,3,0.5);
>> p = Fhi-Flo

p =    0.682689492137086
```

In this problem, `Fhi` provides the probability less than 3.5 and `Flo` provides the probability less than 2.5. The difference of these two probabilities is the probability that x lies between 2.5 and 3.5.

(c) In order to find the probability that a battery lasts more than four years,

$$z_{4.0} = \frac{x - \mu}{\sigma} = \frac{4.0 - 3.0}{.5} = 2.0$$

$$P(4.0 \geq X) = P(2.0 \geq Z) = 1 - P(2.0 \geq Z) = 1 - 0.9772 = 0.0228$$

Alternatively, one can solve the problem in MATLAB.

```
>> p = 1- cdf('normal',4.0,3,0.5)

p =    0.022750131948179
```

In other words, the probability that a battery lasts more than 4.0 years, is 2.28% when the lifetime of this battery is given by a normal distribution with mean 3 years and variance 0.5 years.

5.4. Student's t Distribution

The student's t distribution or simply the t distribution is a generalization of the normal distribution. It is defined as

$$f(t) = \frac{\Gamma[(v+1)/2]}{\Gamma(v/2)\sqrt{\pi v}} \left(1 + \frac{t^2}{v}\right)^{-\frac{v+1}{2}} \quad \text{for} \quad -\infty < t < \infty \tag{5.8}$$

where v is a parameter, called the degrees of freedom, and the gamma function is a standard mathematical function defined as .

$$\Gamma(\alpha) = \int_0^\infty x^{\alpha-1} e^{-x} dx \quad \text{for } \alpha > 0 \tag{5.9}$$

This definite integral cannot be analytically evaluated. The Gamma function has some special values: $\Gamma(1) = 1$, $\Gamma(n) = (n-1)!$ where n is a positive integer, and $\Gamma(1/2) = \pi$. The Gamma function can be evaluated in MATLAB via

```
>> g = gamma(0.5)

g =
1.772453850905516
```

In the limit that the degrees of freedom approaches infinity, the t distribution approaches the normal distribution. One may wonder how this is possible given that the t distribution does not explicitly have the mean and a variance as parameters as does the normal distribution, but we shall save that discussion until the next chapter on sampling.

The t distribution is shown in Figure 5.3. Like the normal distribution it is symmetric about a mean. The random variable t ranges from negative to positive infinity.

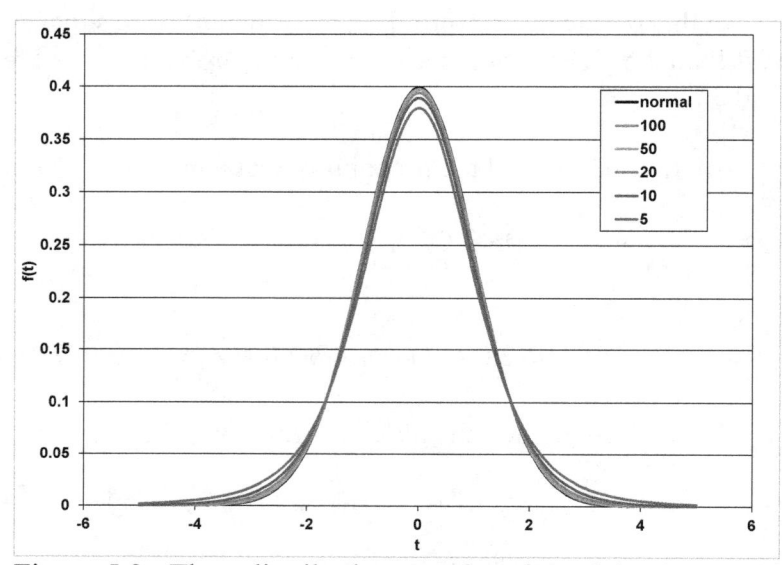

Figure 5.3. The t-distribution as a function of the degrees of freedom and the normal distribution.

When the cavemen were evaluating probabilities from the t distribution with charcoal on the walls of their caves, they used tables of the cumulative t distribution PDF, such as those provided in the back of the statistics textbooks. These tables are also available online. Here we will rely on a modern, computational approach to evaluating the t distribution.

The t Distribution in MATLAB

One can use tools in MATLAB to evaluate problems concerned with the t distribution. In order to generate a random number from the t PDF defined by degrees of freedom, v, one can use the `random` function in MATLAB,

```
>> x = random('t',v)
```

For example, to generate a random number from the t PDF defined by degrees of freedom, 15, one can type the command,

```
>> x = random('t',15)
```

which yields an output such as

```
x =    0.902666218739684
```

In order to evaluate a PDF at a given value of *t*, one can use the `pdf` function in MATLAB

```
>> f = pdf('t',t,v)
```

For example, to generate the value of the PDF for the t PDF defined by degree of freedom, 15, at $t = 0.3$, one can type the command,

```
>> f = pdf('t',0.3,15)
```

which yields the following output

```
f =    0.374018696345112
```

In order to evaluate a cumulative PDF at a given value of *t*, one can use the `cdf` function in MATLAB

```
>> F = cdf('t',thi,v)
```

For example, to generate the value of the cumulative PDF for the t PDF defined by degree of freedom, 15, at $t = 0.3$, one can type the command,

```
>> F = cdf('t',0.3,15)
```

which yields the following output

```
F =    0.615852732732224
```

In other words,

$$F = p(x \leq 0.3) = \int_{-\infty}^{0.3} f(x)dx = 0.61583$$

In order to determine the value of a random variable *t* at which the cumulative PDF of the t distribution has a value of *F*, one can use the `icdf` (the inverse cdf) function in MATLAB

```
>> thi = icdf('t',F,v)
```

For example, to determine the value of a random variable *t* at which the cumulative t PDF, with degrees of freedom of 15, has a value of 0.6, one can type the command,

```
>> thi = icdf('t',0.6,15)
```

which yields the following output

```
thi =    0.257885300937261
```

In other words,

$$F = 0.6 = p(x \leq 0.25789) = \int_{-\infty}^{0.25789} f(x)dx$$

We shall postpone an example of applying the t distribution until the next chapter, where we demonstrate its use in statistical sampling.

5.5. Gamma Distribution

The Gamma distribution commonly appears in the sciences. The Gamma distribution for the continuous random variable *x* with parameters α and β is given by

$$f_\Gamma(x;\alpha,\beta) = \begin{cases} \dfrac{1}{\beta^\alpha \Gamma(\alpha)} x^{\alpha-1} e^{-x/\beta} & \text{for } x > 0 \\ 0 & \text{elsewhere} \end{cases} \qquad (5.10)$$

where the gamma function, defined in equation (5.9) again appears in the distribution.

The Gamma distribution is shown in Figure 5.4. The gamma distribution is only defined for values of *x* greater than zero. Increasing α while holding β constant causes the peak in the distribution to move to larger values of x and causes the magnitude of the peak to decrease. Increasing β while holding α constant also causes the peak in the distribution to move to larger values of x and causes the magnitude of the peak to decrease.

One can use the definitions to explicitly evaluate the mean variance of the random variable *x* in the gamma distribution. Only the results are given here.

$$\mu = \alpha\beta \qquad (5.11)$$

$$\sigma^2 = \alpha\beta^2 \tag{5.12}$$

The Gamma Distribution in MATLAB

One can use tools in MATLAB to evaluate problems concerned with the gamma distribution. In order to generate a random number from the gamma PDF defined by parameters, α and β, one can use the `random` function in MATLAB,

```
>> x = random('gamma',a,b)
```

For example, to generate a random number from the gamma PDF defined by parameters, $\alpha = 2$ and $\beta = 3$, one can type the command,

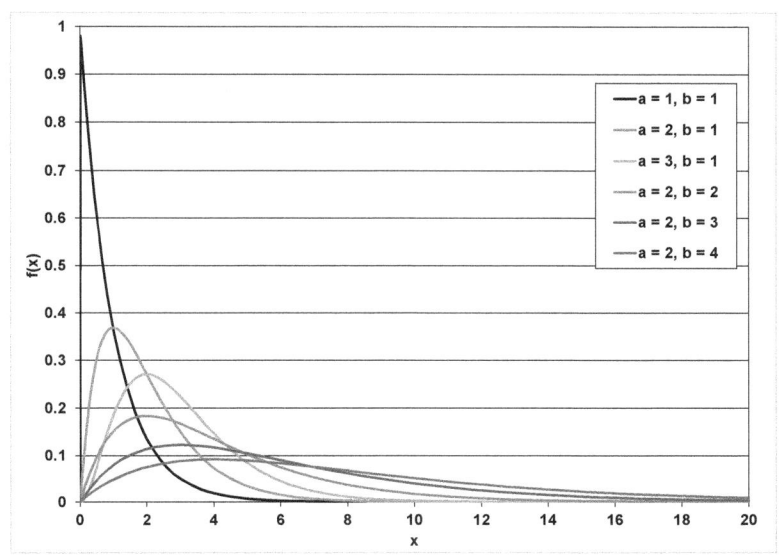

Figure 5.4. The gamma distribution for various values of α and β.

```
>> x = random('gamma',2,3)
```

which yields an output such as

```
x =    8.310470316364883
```

In order to evaluate a PDF at a given value of *x*, one can use the `pdf` function in MATLAB

```
>> f = pdf('gamma',x,a,b)
```

For example, to generate the value of the PDF for the gamma PDF defined by parameters, $\alpha = 2$ and $\beta = 3$, at $x = 1$, one can type the command,

```
>> f = pdf('gamma',1,2,3)
```

which yields the following output

```
f =    0.079614590063754
```

In order to evaluate a cumulative PDF at a given value of x, one can use the `cdf` function in MATLAB

```
>> F = cdf('gamma',xhi,a,b)
```

For example, to generate the value of the cumulative gamma PDF defined by parameters, $\alpha = 2$ and $\beta = 3$, at $x = 5.3$, one can type the command,

```
>> F = cdf('gamma',5.3,2,3)
```

which yields the following output

```
F =    0.527171927915781
```

In other words,

$$F = p(x \leq 5.3) = \int_{-\infty}^{5.3} f(x)dx = 0.52717$$

In order to determine the value of a random variable x at which the cumulative PDF of the gamma distribution has a value of F, one can use the `icdf` (the inverse cdf) function in MATLAB

```
>> xhi = icdf('gamma',F,a,b)
```

For example, to determine the value of a random variable x at which the cumulative gamma PDF, with parameters, $\alpha = 2$ and $\beta = 3$, has a value of 0.6, one can type the command,

```
>> xhi = icdf('gamma',0.6,2,3)
```

which yields the following output

```
xhi =    6.066939735973971
```

In other words,

$$F = 0.6 = p(x \leq 6.06694) = \int_{-\infty}^{6.06694} f(x)dx$$

In an example problem, you more or less have to be told to apply the gamma distribution.

5.6. Exponential Distribution

The exponential distribution is the gamma distribution when α is set to one.. The exponential distribution for the continuous random variable x with parameters β is given by

$$f_e(x;\beta) = \begin{cases} \dfrac{1}{\beta} e^{-x/\beta} & \text{for } x > 0 \\ 0 & \text{elsewhere} \end{cases} \quad (5.13)$$

The exponential distribution is shown in Figure 5.5. The exponential distribution is only defined for values of x greater than zero. Increasing β causes decay of this exponential curve to lengthen.

One can use the definitions to explicitly evaluate the mean variance of the random variable x in the exponential distribution. Only the results are given here.

$$\mu = \beta \quad (5.14)$$

$$\sigma^2 = \beta^2 \quad (5.15)$$

The Exponential Distribution in MATLAB

One can use tools in MATLAB to evaluate problems concerned with the exponential distribution. In order to generate a random number from the exponential PDF defined by parameter, β, one can use the `random` function in MATLAB,

```
>> x = random('exp',b)
```

For example, to generate a random number from the exponential PDF defined by parameter, β =3, one can type the command,

```
>> x = random('exp',3)
```

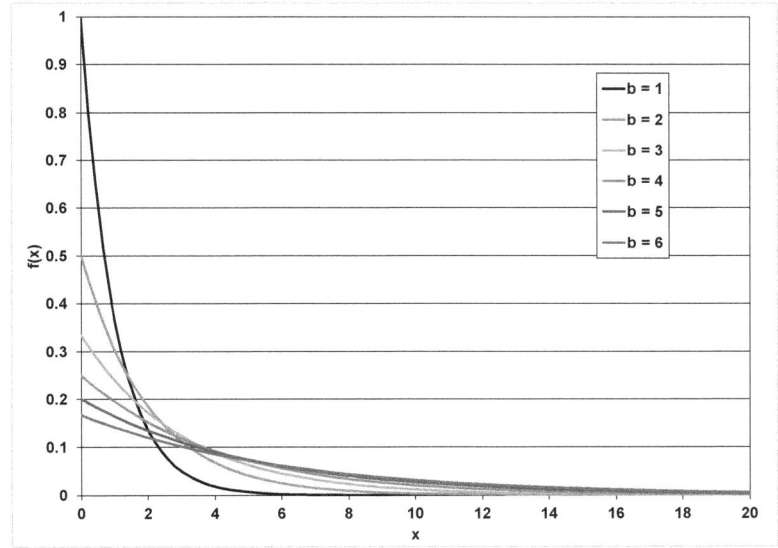

Figure 5.5. The exponential distribution for various values of β.

which yields an output such as

```
x    =    5.858186922301461
```

In order to evaluate a PDF at a given value of x, one can use the `pdf` function in MATLAB

```
>> f = pdf('exp',x,b)
```

For example, to generate the value of the PDF for the exponential PDF defined by parameter, $\beta = 3$, at $x = 1$, one can type the command,

```
>> f = pdf('exp',1,3)
```

which yields the following output

```
f    =    0.238843770191263
```

In order to evaluate a cumulative PDF at a given value of x, one can use the `cdf` function in MATLAB

```
>> F = cdf('exp',xhi,b)
```

For example, to generate the value of the cumulative exponential PDF defined by parameter, $\beta = 3$, at $x = 5.3$, one can type the command,

```
>> F = cdf('exp',5.3,3)
```

which yields the following output

```
F    =    0.829098287198475
```

In other words,

$$F = p(x \leq 5.3) = \int_{-\infty}^{5.3} f(x)dx = 0.82910$$

In order to determine the value of a random variable x at which the cumulative PDF of the exponential distribution has a value of F, one can use the `icdf` (the inverse cdf) function in MATLAB

```
>> xhi = icdf('exp',F,b)
```

For example, to determine the value of a random variable x at which the cumulative exponential PDF, with parameter, $\beta = 3$, has a value of 0.6, one can type the command,

```
>> xhi = icdf('exp',0.6,3)
```

which yields the following output

```
xhi =    3.105378597263349
```

In other words,

$$F = 0.6 = p(x \leq 3.10538) = \int_{-\infty}^{3.10538} f(x)dx$$

Example 5.3: The lifetime of sparkplugs is measure in time, t, and is modelled by the exponential distribution with an average time to failure of 5 years, $\beta = 5$. If new sparkplugs are installed in an 8-cylinder engine and never replaced, what is the probability that m spark plugs are still alive at the end of t_i years where $0 \leq m \leq 8$ and $t_i = 1, 2.5, 5, 7.5, 10$. The sparkplugs are independent of each other.

The probability that a single independent spark plug is still alive at the end of t_i years is given by:

$$P(t_i < t) = \int_{t_i}^{\infty} f_e(t;\beta)dt$$

$$= \int_{t_i}^{\infty} \frac{1}{\beta} e^{-t/\beta} dt = e^{-t_i/\beta}$$

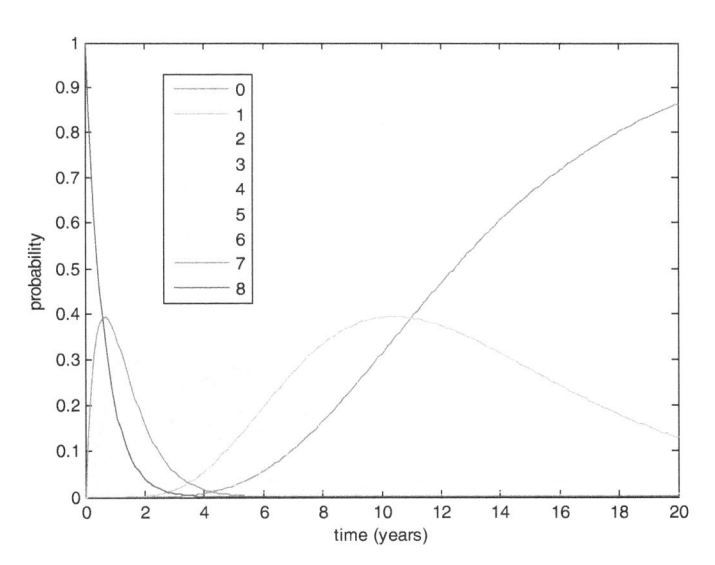

Figure 5.6. The probability that m sparkplugs still work as a function of time, given that the mean sparkplug lifetime is 5 years.

We can compute this for any desired value of t_i. Now, we need that probability that m of $n=8$ sparkplugs are still functioning at t_i, given the probability above. This is precisely the function of the binomial distribution, $b(x;n,p)$ where $x = m$, the number of functioning sparkplugs, $n=8$, the number of total sparkplugs; and $p = P(t_i < t)$, the probability that a single sparkplug makes it to time t_i, as given by the exponential distribution. We can calculate b(m,8, $P(t_i<t)$) for all values of m, (namely $0 \leq m \leq 8$) and for several values of t_i.

In Figure 5.6, w show the probability that m sparkplugs still work as a function of time. Let's examine the plot. At any time t, the sum of the probabilities is 1. At time near 0, it is most probable that all 8 sparkplugs still function. At six years, it is most probable that only 2 sparkplugs still function, followed by 3, 1, 4, 0, 5, 6, 7, 8. At twelve years, it is most probable that no sparkplugs function anymore.

5.7. Chi-Squared Distribution

A second special case of the gamma distribution occurs when α=v/2 and β=2. This distribution is called the chi-squared distribution and v is called the degrees of freedom. The chi-squared distribution for the continuous random variable x with parameter v is given by

$$f_{\chi^2}(x;v) = \begin{cases} \dfrac{1}{2^{v/2}\Gamma(v/2)} x^{v/2-1} e^{-x/2} & \text{for } x > 0 \\ 0 & \text{elsewhere} \end{cases} \quad (5.16)$$

The chi-squared distribution is shown in Figure 5.7. The chi-squared distribution is only defined for values of x greater than zero. Increasing v causes the peak to move out to larger values of the random variable and to decrease in magnitude.

One can use the definitions to explicitly evaluate the mean variance of the random variable x in the chi-squared distribution. Only the results are given here.

$$\mu = v \quad (5.17)$$

$$\sigma^2 = 2v \quad (5.18)$$

The Chi-Squared Distribution in MATLAB

One can use tools in MATLAB to evaluate problems concerned with the chi-squared distribution. In order to generate a random number from the chi-squared PDF defined by parameter, β, one can use the `random` function in MATLAB,

```
>> x = random('chi2',v)
```

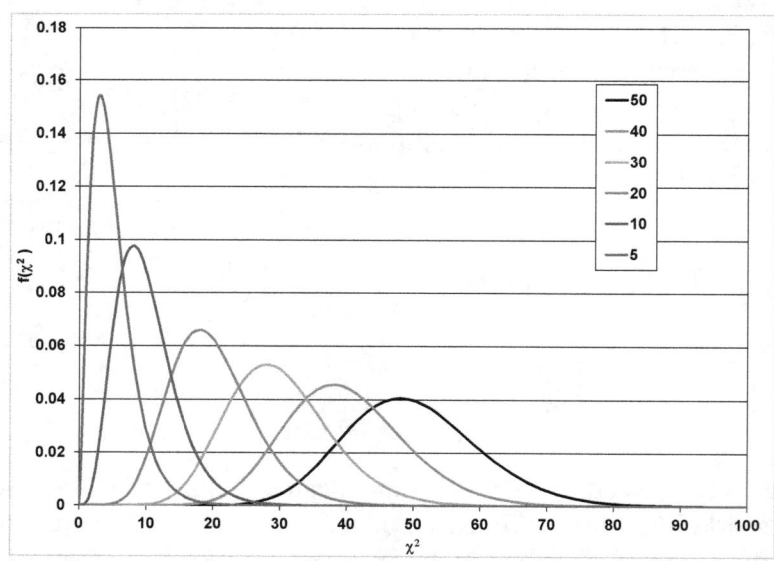

Figure 5.7. The chi-squared distribution for various values of v.

For example, to generate a random number from the chi-squared PDF defined by parameter, v =13, one can type the command,

```
>> x = random('chi2',13)
```

which yields an output such as

```
x =    21.310077394704155
```

In order to evaluate a PDF at a given value of x, one can use the `pdf` function in MATLAB

```
>> f = pdf('chi2',x,v)
```

For example, to generate the value of the PDF for the chi-squared PDF defined by parameter, v=13, at $x = 10$, one can type the command,

```
>> f = pdf('chi2',10,13)
```

which yields the following output

```
f =    0.081773608489854
```

In order to evaluate a cumulative PDF at a given value of x, one can use the `cdf` function in MATLAB

```
>> F = cdf('chi2',xhi,v)
```

For example, to generate the value of the cumulative chi-squared PDF defined by parameter, v = 13, at x = 10, one can type the command,

```
>> F = cdf('chi2',10,13)
```

which yields the following output

```
F =    0.306065632019251
```

In other words,

$$F = p(x \leq 10) = \int_{-\infty}^{10} f(x)dx = 0.30607$$

In order to determine the value of a random variable x at which the cumulative PDF of the chi-squared distribution has a value of F, one can use the `icdf` (the inverse cdf) function in MATLAB

```
>> xhi = icdf('chi2',F,v)
```

For example, to determine the value of a random variable x at which the cumulative chi-squared PDF, with parameter, $v = 13$, has a value of 0.6, one can type the command,

```
>> xhi = icdf('chi2',0.6,13)
```

which yields the following output

```
xhi =    13.635570993661943
```

In other words,

$$F = 0.6 = p(x \leq 13.63557) = \int_{-\infty}^{13.63557} f(x)dx$$

We shall postpone an example of applying the chi-squared distribution until the next chapter, where we demonstrate its use in statistical sampling.

5.8. F distribution

The last distribution of this chapter is the F distribution, which is used in statistical analysis. The F distribution for the continuous random variable x with parameters v_1 and v_2 is given by

$$h_f(f; v_1, v_2) = \begin{cases} \dfrac{\Gamma\left(\dfrac{v_1+v_2}{2}\right)\left(\dfrac{v_1}{v_2}\right)^{\frac{v_1}{2}}}{\Gamma\left(\dfrac{v_1}{2}\right)\Gamma\left(\dfrac{v_2}{2}\right)} \dfrac{f^{\frac{v_1}{2}-1}}{\left(1+\dfrac{v_1}{v_2}f\right)^{\frac{v_1+v_2}{2}}} & \text{for } f > 0 \\ 0 & \text{elsewhere} \end{cases} \quad (5.19)$$

The F distribution is shown in Figure 5.8. The F distribution is only defined for values of x greater than zero. Increasing v_1 while holding v_2 constant causes the peak to move out to larger values of the random variable and to increase in magnitude. Increasing v_2 while holding v_1 constant also causes the peak in the distribution to move out to larger values of the random variable and to increase in magnitude.

The F Distribution in MATLAB

One can use tools in MATLAB to evaluate problems concerned with the F distribution. In order to generate a random number from the F PDF defined by parameter, β, one can use the `random` function in MATLAB,

```
>> x = random('f',v1,v2)
```

For example, to generate a random number from the F PDF defined by parameters, $v_1 = 13$ and $v_2 = 17$, one can type the command,

```
>> x = random('f',13,17)
```

which yields an output such as

```
x =
2.625364477079602
```

In order to evaluate a PDF at a given value of x, one can use the `pdf` function in MATLAB

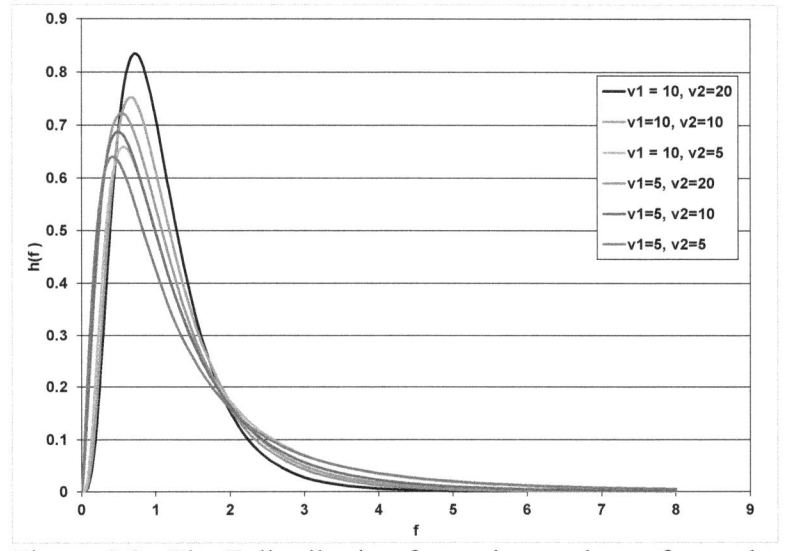

```
>> f = pdf('f',x,v1,v2)
```

Figure 5.8. The F distribution for various values of v_1 and v_2.

For example, to generate the value of the PDF for the F PDF defined by parameters, $v_1 = 13$ and $v_2 = 17$, at $x = 1$, one can type the command,

```
>> f = pdf('f',1,13,17)
```

which yields the following output

```
f =    0.752702954780597
```

In order to evaluate a cumulative PDF at a given value of x, one can use the `cdf` function in MATLAB

```
>> F = cdf('f',xhi,v1,v2)
```

For example, to generate the value of the cumulative F PDF defined by parameters, $v_1 = 13$ and $v_2 = 17$, at x = 1, one can type the command,

```
>> F = cdf('f',1,13,17)
```

which yields the following output

```
F =    0.509323073450741
```

In other words,

$$F = p(x \leq 1) = \int_{-\infty}^{10} f(x)dx = 0.50932$$

In order to determine the value of a random variable x at which the cumulative PDF of the F distribution has a value of F, one can use the `icdf` (the inverse cdf) function in MATLAB

```
>> xhi = icdf('f',F,v1,v2)
```

For example, to determine the value of a random variable x at which the cumulative F PDF, with parameters, $v_1 = 13$ and $v_2 = 17$, has a value of 0.6, one can type the command,

```
>> xhi = icdf('f',0.6,13,17)
```

which yields the following output

```
xhi =    1.129264291748036
```

In other words,

$$F = 0.6 = p(x \leq 1.12926) = \int_{-\infty}^{1.12926} f(x)dx$$

We shall postpone an example of applying the F distribution until the next chapter, where we demonstrate its use in statistical sampling.

5.9. Functions of Random Variables

Any PDF of a variable can be transformed into a PDF of a function of that variable. In general, $f(x)$ can be transformed to $g(y(x))$, where $f(x)$ is the PDF describing the distribution of the random variable x and $g(y(x))$ is the PDF describing the distribution of $y(x)$.

Continuous Probability Distributions - 129

Example 5.4: When we changed the variable of a normal distribution, $f(x;\mu,\sigma)$, to the standard normal distribution, $f(z;\mu=0,\sigma=1)$, we used the transformation,

$$z(x) = \frac{x-\mu}{\sigma}$$

Thus $f(x;\mu,\sigma)$ is the PDF describing the distribution of the random variable x and $f(z;\mu=0,\sigma=1)$ is the PDF describing the distribution of the random variable $z(x)$.

Example 5.5: Let x be a continuous random variable with PDF

$$f(x) = \begin{cases} \dfrac{x}{12} & \text{for } 1 < x < 5 \\ 0 & \text{elsewhere} \end{cases}$$

then the probability, P(x<a) is given by

$$P(x<a) = \int_1^a f(x')dx' = \int_1^a \frac{x'}{12}dx' = \frac{a^2-1}{24}$$

We can also find the probability distribution for y, where $y(x) = 2x+2$. Rearranging for x yields $x = \dfrac{y-2}{2}$ and $dx = \dfrac{dy}{2}$

$$g(y) = f((y-2)/2) = \frac{(y-2)/2}{12} = \frac{y-2}{24}$$

with limits $y(x=1)=4$ and $y(x=5)=12$. Therefore,

$$P(x<a) = P(y(x)<y(a)) = \int_1^a f(x')dx' = \int_4^{y(a)} g(y')\frac{dy'}{2}$$

$$P(y(x)<y(a)) = \int_4^{y(a)} \frac{y'-2}{24}\frac{dy'}{2} = \left(\frac{y'^2}{96} - \frac{4y'}{96}\right)\Bigg|_4^{y(a)} = \frac{y(a)^2 - 4y(a)}{96}$$

In Figure 5.9, we show f(x) and F(x)=P(x<X), the cumulative PDF, as a function of x, and g(y) and G(y)=P(y<Y), the cumulative PDF, as a function of y.

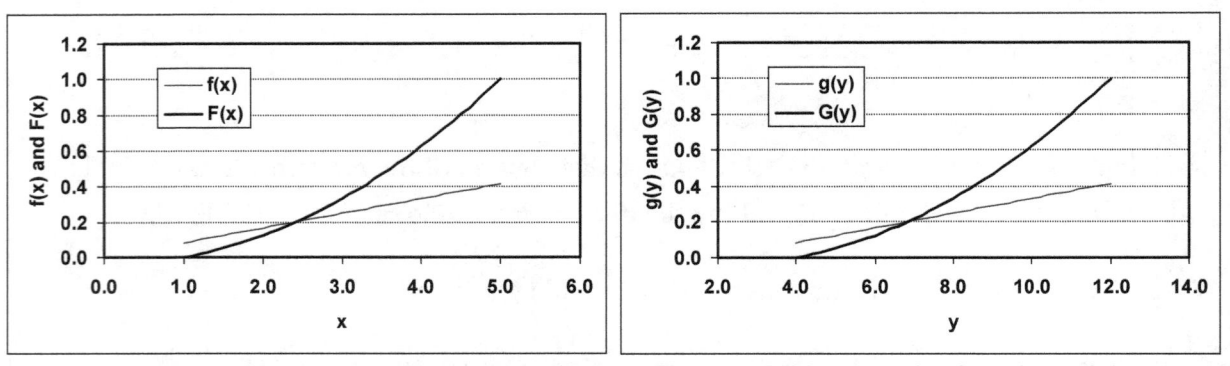

Figure 5.9. The PDF and cumulative PDF for a random variable, x, and a function of the random variable, $y(x) = 2x + 2$

Example 5.6: Consider the continuous random variable describing the temperature in a furnace, T, which is distributed between 500 K and 1000 K according to a PDF is given by $f(T) = c/T^2$, where $c = 1000$. The diffusivity is given as a function of temperature, $D(T) = D_o \exp\left(-E_a/RT\right)$ cm²/s, where $D_o = 5$ cm²/s and $E_a = 800R$. Construct a PDF in terms of the diffusivity rather than the temperature. Plot the PDF and CDF for both temperature and diffusivity.

We can invert the relationship between D and T to solve for the temperature.

$$T = \frac{E_a}{R} \frac{1}{\ln\left(\frac{D_o}{D}\right)}$$

We can substitute this into the PDF for T to obtain the functional form of the PDF for D.

$$f(D) = c/T^2 = c\left[\frac{R}{E_a} \ln\left(\frac{D_o}{D}\right)\right]^2$$

The PDF in T was non-zero in the range 500 < T < 1000. The PDF in D is non-zero in the range D(500) < D < D(1000) or approximately, 1.009 < D < 2.247.

Plots of both the PDF in temperature and the PDF in diffusivity are shown in Figure 5.10. The CDF is also plotted.

The cumulative PDF in T is given by

$$F(T) = \int_{-\infty}^{T} f(T) dT = \int_{-\infty}^{T} c/T^2 \, dT = -c\left[\frac{1}{T}\right]_{T_{low}}^{T} = \frac{c}{T_{low}} - \frac{c}{T}$$

The cumulative PDF in D is obtained through substitution and is given by

$$F(D) = c \frac{R}{E_a} \ln\left(\frac{D(T)}{D(T_{low})}\right)$$

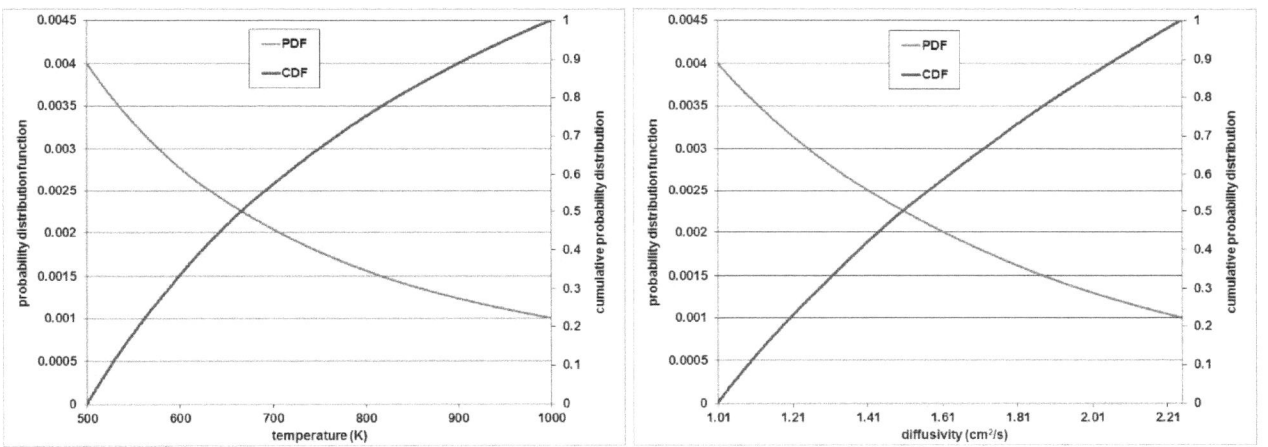

Figure 5.10. The PDF and CDF for the temperature (left) and the diffusivity (right). The curves are not exactly the same shape because the relationship was between temperature and diffusivity is nonlinear.

5.10. Problems

Problem 5.1.
Given a standard normal distribution, find the area under the curve which lies
(a) to the left of z = 1.43
(b) to the right of z = -0.89
(c) between z = -2.16 and z = -0.65

Problem 5.2.
The daily amount of coffee, in liters, dispensed by a machine located in an airport lobby is a random variable X having a continuous uniform distribution with A = 7 and B = 10. Find the probability that on a given day the amount of coffee dispensed by this machine will be
(a) at most 8.8 liters
(b) more than 7.4 liters but less than 9.5 liters
(c) at least 8.5 liters.

Problem 5.3.
In the November 1990 issue of Chemical Engineering Progress a study discussed the percent purity of oxygen from a certain supplier. Assume that the mean was 99.61 with a standard deviation of 0.08. Assume that the distribution of percent purity was approximately normal.
   (a) What percentage of the purity values would you expect to be between 99.5 and 99.7?
   (b) What purity value would you expect to exceed exactly 5% of the population?

Problem 5.4.
The average life of a certain type of small motor is 10 years with a standard deviation of two years. The manufacturer replaces all free all motors that fail while under guarantee. If he is willing to replace only 3% of the motors that fail, how long a guarantee should he offer? Assume that the lifetime of a motor follows the normal distribution.

Problem 5.5.
   You flip a coin 100 times.
   (a) What is the probability that you obtain at least 40 heads? Use the binomial distribution.
   (b) What is the probability that you obtain at least 40 heads? Use the normal approximation to the binomial distribution.

Problem 5.6.
   The life of a certain type of device has an advertised failure rate or 0.01 per hour. The failure rate is constant and the exponential distribution applies.
   (a) What is the mean time to failure?
   (b) What is the probability that 200 hours will pass before a failure is observed?
   (c) What is the probability that exactly 3 of 5 devices are functioning after 200 hours?

Problem 5.7.
   Using the gamma distribution with $\alpha = 4$ and $\beta = 1/3$, find
   (a) $P(x < 1/2)$
   (b) C where $P(x < C) = 0.05$

Problem 5.8.
   Using the Chi-Squared distribution with $v=10$, find
   (a) $P(\chi^2 < 4)$
   (b) C where $P(\chi^2 < C) = 0.05$

Problem 5.9.
   Using the t-distribution with $v=10$, find
   (a) $P(t < -1)$
   (b) C where $P(t < C) = 0.05$

Problem 5.10.
Using the F-distribution with $v_1=8$ and $v_2 = 12$, find
(a) $P(F<1)$
(b) C where $P(F<C) = 0.05$

Chapter 6. Sampling and Estimation

6.1. Introduction

Frequently the engineer is unable to completely characterize the entire population. She/he must be satisfied with examining some subset of the population, or several subsets of the population, in order to infer information about the entire population. Such subsets are called **samples**. A **population** is the entirety of observations and a sample is a subset of the population. A sample that gives correct inferences about the population is a **random sample**, otherwise it is **biased**.

Statistics are given different symbols than the expectation values because *statistics are approximations of the expectation value*. The statistic called the mean is an approximation to the expectation value of the mean. The statistic mean is the mean of the sample and the expectation value mean is the mean of the entire population. In order to calculate an expectation, one requires knowledge of the PDF. In practice, the motivation in calculating a statistic is that one has no knowledge of the underlying PDF.

6.2. Statistics

Any function of the random variables constituting a random sample is called a statistic.

Example 6.1.: **Mean**
The mean is a statistic of a random sample of size n and is defined as

$$\overline{X} = \frac{1}{n}\sum_{i=1}^{n} X_i \tag{6.1}$$

Example 6.2.: **Median**
The median is a statistic of a random sample of size n, which represents the "middle" value of the sample and, for a sampling arranged in increasing order of magnitude, is defined as

$$\widetilde{X} = X_{(n+1)/2} \quad \text{for odd n}$$

$$\widetilde{X} = \frac{X_{n/2} + X_{(n+1)/2}}{2} \quad \text{for even n} \tag{6.2}$$

The median of the sample space {1,2,3} is 2.
The median of the sample space {3,1,2} is 2.
The median of the sample space {1,2,3,4} is 2.5.

Example 6.3.: **Mode**
  The mode is a statistic of a random sample of size n, which represents the most frequently appearing value in the sample. The mode may not exist and, if it does, it may not be unique.

The mode of the sample space {2,1,2,3} is 2.
The mode of the sample space {2,1,2,3,4,4} is 2 and 4. (bimodal)
The mode of the sample space {1,2,3} does not exist since each entry occurs only once.

Example 6.4.: **Range**
  The range is a statistic of a random sample of size n, which represents the "span" of the sample and, for a sampling arranged in increasing order of magnitude, is defined as

$$range(X) = X_n - X_1 \tag{6.3}$$

The range of {1,2,3,4,5} is 5-1=4.

Example 6.5.: **Variance**
  The variance is a statistic of a random sample of size n, which represents the "spread" of the sample and is defined as

$$S^2 = \frac{\sum_{i=1}^{n}(X_i - \overline{X})^2}{n-1} = \frac{n\sum_{i=1}^{n}(X_i)^2 - \left(\sum_{i=1}^{n}X_i\right)^2}{n(n-1)} \tag{6.4}$$

The reason for using (n-1) in the denominator rather than n is given later.

Example 6.6.: **Standard Deviation**
  The standard deviation, s, is a statistic of a random sample of size n, which represents the "spread" of the sample and is defined as the positive square root of the variance.

$$S = \sqrt{S^2} \tag{6.5}$$

6.3. Sampling Distributions

We have now stated the definitions of the statistics we are interested in. Now, we need to know the distribution of the statistics to determine how good these sampling approximations are to the true expectation values of the population.

Statistic 1. Mean when the variance is known: Sampling Distribution

If $\bar{X}$ is the mean of a random sample of size n taken from a population with mean μ and variance σ^2, then the limiting form of the distribution of

$$Z = \frac{\bar{X} - \mu}{\sigma/\sqrt{n}} \tag{6.6}$$

as $n \to \infty$, is the **standard normal distribution** $n(z;0,1)$. This is known as the Central Limit Theorem. What this says is that, given a collection of random samples, each of size n, yielding a mean $\bar{X}$, the distribution of $\bar{X}$ approximates a normal distribution, and becomes exactly a normal distribution as the sample size goes to infinity. The distribution of X does not have to be normal. Generally, the normal approximation for $\bar{X}$ is good if $n > 30$.

We provide a derivation in Appendix V proving that the distribution of the sample mean is given by the normal distribution.

Example 6.7.: distribution of the mean, variance known
In a reactor intended to grow crystals in solution, a "seed" is used to encourage nucleation. Individual crystals are randomly sampled from the effluent of each reactor of sizes $n = 10$. The population has variance in crystal size of $\sigma^2 = 1.0 \ \mu m^2$. (We must know this from previous research.) The samples yield mean crystal sizes of $\bar{x} = 15.0 \ \mu m$. What is the likelihood that the true population mean, μ, is actually less than 14.0 μm?

$$Z = \frac{\bar{x} - \mu}{\sigma/\sqrt{n}} = \frac{15 - 14}{1/\sqrt{10}} = 3.162$$

$$P(\mu < 14) = P(z > 3.162)$$

We have the change in sign because as μ increases, z decreases.

The evaluation of the cumulative normal probability distribution can be performed several ways. First, when the pioneers were crossing the plains in their covered wagons and they wanted to evaluate probabilities from the normal distribution, they used Tables of the cumulative normal PDF, such as those provided in the back of the statistics textbook. These tables are also available online. For example wikipedia has a table of cumulative standard numeral PDFs at http://en.wikipedia.org/wiki/Standard_normal_table

Using the table, we find

$$P(\mu<14)=P(z>3.162)=1-P(z<3.162)=1-0.9992=0.0008$$

Second, we can use a modern computational tool like MATLAB to evaluate the probability. The problem can be worked in terms of the standard normal PDF (μ = 0 and σ = 1), which for $P(\mu<14)=P(z>3.162)=1-P(z<3.162)$ is

```
>> p = 1 - cdf('normal',3.162,0,1)

p = 7.834478217108032e-04
```

Alternatively, the problem can be worked in terms of the non-standard normal PDF ($\bar{x}=15$ and $\sigma/\sqrt{n}=1/\sqrt{10}$), which for $P(\mu<14)$

```
>> p = cdf('normal',14,15,1/sqrt(10))

p = 7.827011290012762e-04
```

The difference in these results is due to the round-off in 3.162, used as an argument in the function call for the standard normal distribution.

Based on our sampling data, the probability that the true sample mean is less than 14.0 μm is 0.078%.

Statistic 2. difference of means when the variance is known: Sampling Distribution

It is useful to know the sampling difference of two means when you want to determine whether there is a significant difference between two populations. This situation applies when you takes two random samples of size n_1 and n_2 from two different populations, with means μ_1 and μ_2 and variances σ_1^2 and σ_2^2, respectively. Then the sampling distribution of the difference of means, $\bar{X}_1 - \bar{X}_2$, is approximately normal, distributed with mean

$$\mu_{\bar{X}_1-\bar{X}_2} = \mu_1 - \mu_2$$

and variance

$$\sigma_{\bar{X}_1-\bar{X}_2}^2 = \frac{\sigma_1^2}{n_1} - \frac{\sigma_2^2}{n_2}$$

Hence,

$$Z = \frac{(\bar{X}_1 - \bar{X}_2) - (\mu_1 - \mu_2)}{\sqrt{\left(\frac{\sigma_1^2}{n_1}\right) + \left(\frac{\sigma_2^2}{n_2}\right)}} \tag{6.7}$$

is approximately a standard normal variable.

Example 6.8.: distribution of the difference of means, variances known
In a reactor intended to grow crystals, two different types of "seeds" are used to encourage nucleation. Individual crystals are randomly sampled from the effluent of each reactor of sizes $n_1 = 10$ and $n_2 = 20$. The populations have variances in crystal size of $\sigma_1^2 = 1.0$ μm² and $\sigma_2^2 = 2.0$ μm². (We must know this from previous research.) The samples yield mean crystal sizes of $\bar{X}_1 = 15.0$ μm and $\bar{X}_2 = 10.0$ μm. How confident can we be that the true difference in population means, $\mu_1 - \mu_2$, is actually 4.0 μm or greater?
Using equation (6.7) we have:

$$Z = \frac{(\bar{X}_1 - \bar{X}_2) - (\mu_1 - \mu_2)}{\sqrt{\left(\frac{\sigma_1^2}{n_1}\right) + \left(\frac{\sigma_2^2}{n_2}\right)}} = \frac{(15-10) - (4)}{\sqrt{\left(\frac{1}{10}\right) + \left(\frac{2}{20}\right)}} = 2.2361$$

$$P(\mu_1 - \mu_2 > 4.0) = P(z < 2.2361)$$

We have the change in sign because as $\Delta\mu$ increases, z decreases. The probability that $\mu_1 - \mu_2$ is greater 4.0 μm is then given by $P(Z<2.2361)$. How do we know that we want $P(Z<2.2361)$ and not $P(Z>2.2361)$? We just have to sit down and think what the problem physically means. Since we want the probability that $\mu_1 - \mu_2$ is greater 4.0 μm, we know we need to include the area due to higher values of $\mu_1 - \mu_2$. Higher values of $\mu_1 - \mu_2$ yield lower values of Z. Therefore, we need the less than sign.
The evaluation of the cumulative normal probability distribution can again be performed two ways. First, using a standard normal table, we have

$P(Z < 2.24) = 0.9875$

Second, using MATLAB we have

```
>> p = cdf('normal',2.2361,0,1)
```

```
p =   0.987327389270190
```

We expect 98.73% of the differences in crystal size of the two populations to be at least 4.0 µm.

Statistic 3. Mean when the variance is unknown: Sampling Distribution

Of course, usually we don't know the population variance. In that case, we have to use some other statistic to get a handle on the distribution of the mean.

If $\overline{X}$ is the mean of a random sample of size n taken from a population with mean μ and unknown variance, then the limiting form of the distribution of

$$T = \frac{\overline{X} - \mu}{S/\sqrt{n}} \tag{6.8}$$

as $n \to \infty$, is the **t distribution** $f_T(t;v)$. The T-statistic has a t-distribution with $v = n-1$ degrees of freedom. The t-distribution is just another continuous PDF, like the others we learned about in the previous section.

The t distribution is given by

$$f(t) = \frac{\Gamma[(v+1)/2]}{\Gamma(v/2)\sqrt{\pi v}} \left(1 + \frac{t^2}{v}\right)^{-\frac{v+1}{2}} \quad \text{for} \quad -\infty < t < \infty$$

As a reminder, the t distribution is plotted again in Figure 6.1.

Example 6.9.: distribution of the mean, variance unknown

In a reactor intended to grow crystals, a "seed" is used to encourage nucleation. Individual crystals are randomly sampled from the effluent of each reactor of sizes $n = 10$. The population has unknown variance in crystal size. The samples yield mean crystal sizes of $\overline{x} = 15.0$ µm and a

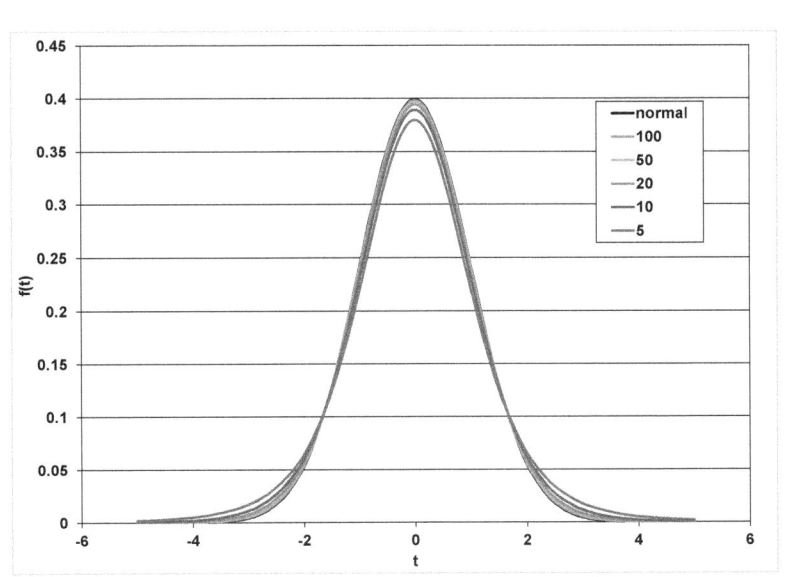

Figure 6.1. The t distribution as a function of the degrees of freedom and the normal distribution.

sample variance of $s^2 = 1.0 \, \mu m^2$. What is the likelihood that the true population mean, μ, is actually less than 14.0 µm?

$$t = \frac{\bar{x} - \mu}{s/\sqrt{n}} = \frac{15-14}{1/\sqrt{10}} = 3.162$$

$$P(\mu < 14) = P(t > 3.162)$$

We have the change in sign because as μ increases, t decreases. The parameter $v = n-1 = 9$.

The evaluation of the cumulative t probability distribution can again be performed two ways. First, we can use a table of critical values of the t-distribution. It is crucial to note that such a table does not provide cumulative PDFs, rather it provides one minus the cumulative PDF. In other words, where as the standard normal table provides the probability less than z (the cumulative PDF), the t-distribution table provides the probability greater than t (one minus the cumulative PDF). We then have

$$P(\mu < 14) = P(t > 3.162) \approx 0.007$$

Second, using MATLAB we have $P(\mu < 14) = P(t > 3.162) = 1 - P(t < 3.162)$

```
>> p = 1 - cdf('t',3.162,9)
p =   0.005756562560207
```

Based on our sampling data, the probability that the true sample mean is less than 14.0 µm is 0.57%.

We should point out that our percentage here is substantially greater than for our percentage when we knew the population variance (0.078%). That is because knowing the population variance reduces our uncertainty. Approximating the population variance with the sampling variance adds to the uncertainty and results in a larger percentage of our population deviating farther from the sample mean.

Example 6.10.:   distribution of the mean, variance unknown
An engineer claims that the population mean yield of a batch process is 500 g/ml of raw material. To verify this, she samples 25 batches each month. One month the sample has a mean $\bar{X} = 518$ g and a standard deviation of s=40 g. Does this sample support his claim that $\mu = 500$ g?

The first step in solving this problem is to compute the T statistic.

Sampling and Estimation - 141

$$T = \frac{\overline{X} - \mu}{S/\sqrt{n}} = \frac{500 - 518}{40/\sqrt{25}} = -2.25$$

Second, using MATLAB we have $P(\mu > 518) = P(t < -2.25)$

```
>> p = cdf('t',-2.25,24)

p =   0.016944255452754
```

(Or using a Table, we find that when $v=24$ and $T=2.25$, $\alpha=0.02$). This means there is only a 1.6% probability that a population with $\mu = 500$ would yield a sample with $\overline{X} = 518$ or higher. Therefore, it is unlikely that 500 is the population mean.

Statistic 4. difference of means when the variance is unknown: Sampling Distribution
It is useful to know the sampling difference of two means when you want to determine whether there is a significant difference between two populations. Sometimes you want to do this when you don't know the population variances. This situation applies when you takes two random samples of size n_1 and n_2 from two different populations, with means μ_1 and μ_2 and unknown variances. Then the sampling distribution of the difference of means, $\overline{X}_1 - \overline{X}_2$, follows the t-distribution.

transformation: $$T = \frac{(\overline{X}_1 - \overline{X}_2) - (\mu_1 - \mu_2)}{\sqrt{\left(\frac{s_1^2}{n_1}\right) + \left(\frac{s_2^2}{n_2}\right)}} \qquad (6.9)$$

symmetry: $t_{1-\alpha} = -t_\alpha$,

parameters: $v = n_1 + n_2 - 2$ if $\sigma_1 = \sigma_2$

parameters: $$v = \frac{\left(\frac{s_1^2}{n_1} + \frac{s_2^2}{n_2}\right)^2}{\left[\left(\frac{s_1^2}{n_1}\right)^2 \bigg/ (n_1 - 1)\right] + \left[\left(\frac{s_2^2}{n_2}\right)^2 \bigg/ (n_2 - 1)\right]} \quad \text{if } \sigma_1 \neq \sigma_2$$

Since we don't know either population variance in this case, we can't assume they are equal unless we are told they are equal.

Example 6.11.: distribution of the difference of means, variances unknown
In a reactor intended to grow crystals, two different types of "seeds" are used to encourage nucleation. Individual crystals are randomly sampled from the effluent of each reactor of sizes

$n_1 = 10$ and $n_2 = 20$. The populations have unknown variances in crystal size. The samples yield mean crystal sizes of $\bar{X}_1 = 15.0$ μm and $\bar{X}_2 = 10.0$ μm and sample variances of $s_1^2 = 1.0$ μm² and $s_2^2 = 2.0$ μm². What percentage of true population differences yielding these sampling results would have a true difference in population means, $\mu_1 - \mu_2$, of 4.0 μm or greater?

$$T = \frac{(\bar{X}_1 - \bar{X}_2) - (\mu_1 - \mu_2)}{\sqrt{\left(\frac{s_1^2}{n_1}\right) + \left(\frac{s_2^2}{n_2}\right)}} = \frac{(15-10) - (4)}{\sqrt{\left(\frac{1}{10}\right) + \left(\frac{2}{20}\right)}} = 2.2361$$

The degree of freedom parameter is given by:

$$v = \frac{\left(\frac{s_1^2}{n_1} + \frac{s_2^2}{n_2}\right)^2}{\left[\left(\frac{s_1^2}{n_1}\right)^2 / (n_1 - 1)\right] + \left[\left(\frac{s_2^2}{n_2}\right)^2 / (n_2 - 1)\right]} = \frac{\left(\frac{1^2}{10} + \frac{2^2}{20}\right)^2}{\left[\left(\frac{1^2}{10}\right)^2 / (10-1)\right] + \left[\left(\frac{2^2}{20}\right)^2 / (20-1)\right]} = 27.98 \approx 28$$

$$P(\mu_1 - \mu_2 > 4.0) = P(t < 2.2361) = 1 - P(t > 2.2361)$$

The evaluation of the cumulative normal probability distribution can again be performed two ways. First, using a table of critical values of the t-distribution, we have

$$P(\mu_1 - \mu_2 > 4.0) = P(t < 2.2361) = 1 - P(t > 2.2361) = 1 - 0.0217 = 0.9783$$

Second, using MATLAB we have for $P(\mu_1 - \mu_2 > 4.0) = P(t < 2.2361)$

```
>> p = cdf('t',2.2361,28)

p =   0.983252747598848
```

We expect 98.3% of the differences in crystal size of the two populations to be at least 4.0 μm.

Statistic 5. Variance: Sampling Distribution

We now wish to know the sampling distribution of the sample variance, S^2. If S^2 is the variance of a random sample of size n taken from a population with mean μ and variance σ^2, then the statistic

$$\chi^2 = \frac{(n-1)S^2}{\sigma^2} = \sum_{i=1}^{n} \frac{(X_i - \overline{X})^2}{\sigma^2} \qquad (6.10)$$

has a chi-squared distribution with $v=n-1$ degrees of freedom, $f_{\chi^2}(\chi^2; n-1)$. The chi-squared distribution is defined as

$$f_{\chi^2}(x;v) = \begin{cases} \dfrac{1}{2^{v/2}\Gamma(v/2)} x^{v/2-1} e^{-x/2} & \text{for } x > 0 \\ 0 & \text{elsewhere} \end{cases}$$

It is a special case of the Gamma Distribution, when $\alpha=v/2$ and $\beta=2$, where v is called the "degrees of freedom" and is a positive integer. As a reminder, we provide a plot of the chi-squared distribution in Figure 6.2.

Example 6.12.: distribution of the variance

In a reactor intended to grow crystals, a "seed" is used to encourage nucleation. Individual crystals are randomly sampled from the effluent of each reactor of sizes $n=10$. The samples yield mean crystal sizes of

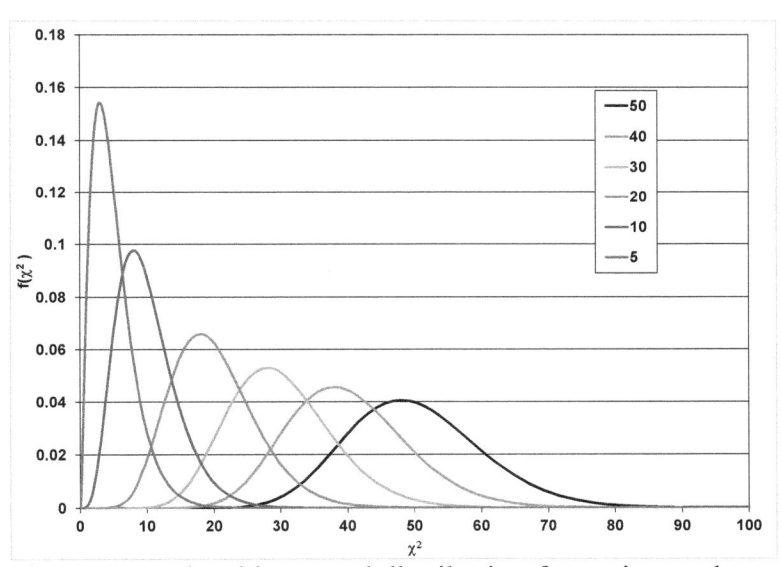

Figure 6.2. The chi-squared distribution for various values of v.

$\overline{x}=15.0$ μm and a sample variance of $s^2 = 1.0\,\mu m^2$. What is the likelihood that the true population variance, σ^2, is actually less than 0.5 μm²?

$$\chi^2 = \frac{(n-1)S^2}{\sigma^2} = \frac{(10-1)1}{0.5} = 18$$

$$P(\sigma^2 < 0.5) = P(\chi^2 > 18)$$

We have the change in sign because as σ^2 increases, χ^2 decreases. The parameter $v = n-1 = 9$.

The evaluation of the cumulative χ^2 probability distribution can again be performed two ways. First, we can use a table of critical values of the χ^2-distribution. It is crucial to note that such a table does not provide cumulative PDFs, rather it provides one minus the cumulative PDF. We then have

$$P(\sigma^2 < 0.5) = P(\chi^2 > 18) \approx 0.04$$

Second, using MATLAB we have $P(\sigma^2 < 0.5) = P(\chi^2 > 18) = 1 - P(\chi^2 < 18)$

```
>> p = 1 - cdf('chi2',18,9)

p =    0.035173539466985
```

Based on our sampling data, the probability that the true variance is less than 0.5 μm² is 3.5%.

Statistic 6. the ratio of 2 Variances: Sampling Distribution (F-distribution)

Just as we studied the distribution of two sample means, so too are we interested in the distribution of two variances. In the case of the mean, it was a difference. In the case of the variance, the ratio is more useful. Now consider sampling two random samples of size n_1 and n_2 from two different populations, with means σ_1^2 and σ_2^2, respectively. The statistic, F,

$$F = \frac{S_1^2 / \sigma_1^2}{S_2^2 / \sigma_2^2} = \frac{S_1^2 \sigma_2^2}{S_2^2 \sigma_1^2} \tag{6.11}$$

provides a distribution of the ratio of two variances. This distribution is called the F-distribution with $v_1 = n_1 - 1$ and $v_2 = n_2 - 1$ degrees of freedom. The f-distribution is defined as

$$h_f(f; v_1, v_2) = \begin{cases} \dfrac{\Gamma\left(\dfrac{v_1+v_2}{2}\right)\left(\dfrac{v_1}{v_2}\right)^{\frac{v_1}{2}}}{\Gamma\left(\dfrac{v_1}{2}\right)\Gamma\left(\dfrac{v_2}{2}\right)} \dfrac{f^{\frac{v_1}{2}-1}}{\left(1+\dfrac{v_1}{v_2}f\right)^{\frac{v_1+v_2}{2}}} & \text{for } f > 0 \\ 0 & \text{elsewhere} \end{cases}$$

As a reminder, the f-distribution is plotted in Figure 6.3.

Example 6.13.: ratio of the variances

In a reactor intended to grow crystals, two different types of "seeds" are used to encourage nucleation. Individual crystals are randomly sampled from the effluent of each reactor of sizes

$n_1 = 10$ and $n_2 = 20$. The populations have unknown variances in crystal size. The samples yield mean crystal sizes of $\overline{X}_1 = 15.0$ μm and $\overline{X}_2 = 10.0$ μm and sample variances of $s_1^2 = 1.0$ μm² and $s_2^2 = 2.0$ μm². What is the probability that the ratio of variances, $\dfrac{\sigma_1^2}{\sigma_2^2}$, is less than 0.25?

$$F = \frac{S_1^2 \sigma_2^2}{S_2^2 \sigma_1^2} = \frac{1}{2 \cdot 0.25} = 2$$

$$P\left(\frac{\sigma_1^2}{\sigma_2^2} < 0.25\right) = P(F > 2)$$

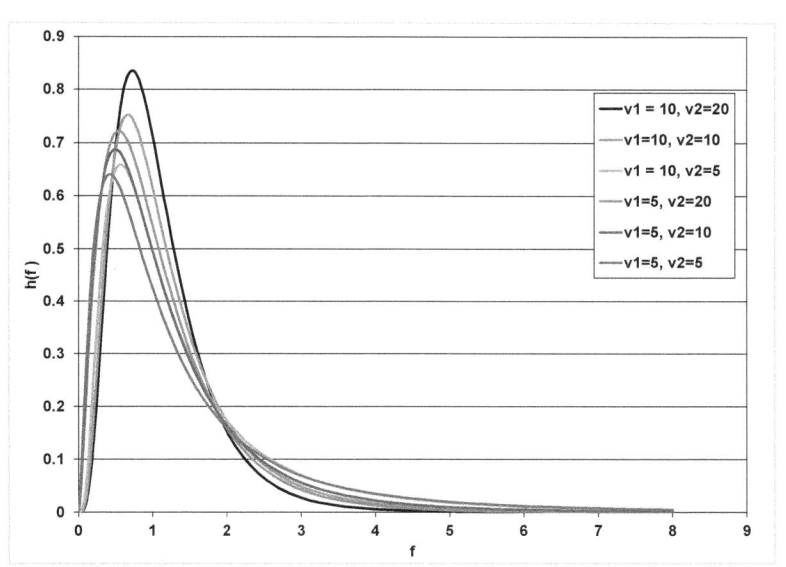

Figure 6.3. The F distribution for various values of v_1 and v_2.

We have the change in sign because as $\dfrac{\sigma_1^2}{\sigma_2^2}$ increases, F decreases. The parameters are $v_1 = n_1 - 1 = 9$ and $v_2 = n_2 - 1 = 19$.

The evaluation of the cumulative F probability distribution can again be performed in one way. We cannot use tables because there are no tables for arbitrary values of the probability. There are only tables for two values of the probability, 0.01 and 0.05. Therefore, using MATLAB we have $P\left(\dfrac{\sigma_1^2}{\sigma_2^2} < 0.25\right) = P(F > 2) = 1 - P(F < 2)$

```
>> p = 1 - cdf('f',2,9,19)
p =  0.097413204997132
```

Based on our sampling data, the probability that the ratio of variances is less than 0.25 is 9.7%.

6.4. Confidence Intervals

In the previous section we showed what types of distributions describe various statistics of a random sample. In this section, we discuss estimating the population mean and variance from the sample mean and variance. In addition, we introduce confidence intervals to quantify the goodness of these estimates.

A confidence interval is some subset of random variable space with which someone can say something like, "I am 95% sure that the true population mean is between μ_{low} and μ_{hi}." In this section, we discuss how a confidence interval is defined and calculated.

The confidence interval is defined by a percent. This percent is called (1-2α). So if α=0.05, then you would have a 90% confidence interval.

The concept of a confidence interval is illustrated in graphical terms in Figure 6.4.

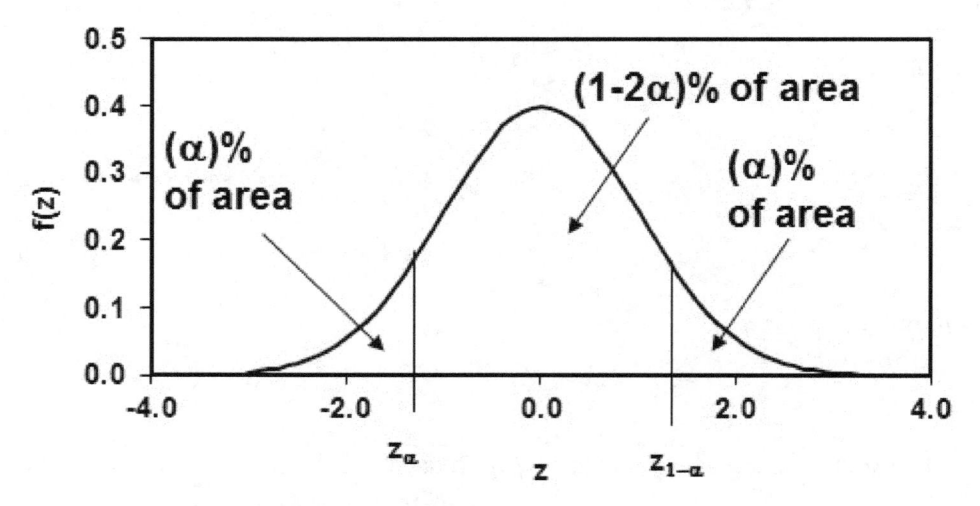

Figure 6.4. A schematic illustrating a confidence interval.

The trick then is to find $\mu_{low} = z_\alpha$ and $\mu_{hi} == z_{1-\alpha}$ so that you can say for a given α, I am $(1 - 2\alpha)\%$ confident that $\mu_{low} < \mu < \mu_{hi}$.

Statistic 1. mean, σ known: confidence interval

We now know that the sample mean is distributed with the standard normal distribution. For a symmetric PDF, centered around zero, like the standard normal, $\mu_{low} = -\mu_{hi}$. We can then make the statement:

$$P(z_\alpha < Z < z_{1-\alpha}) = 1 - 2\alpha$$

Now the normal distribution is symmetric about the y-axis so we can write

$$z_\alpha = -z_{1-\alpha}$$

so

Sampling and Estimation - 147

$$P(z_\alpha < Z < z_{1-\alpha}) = P(z_\alpha < Z < -z_\alpha) = 1 - 2\alpha$$

where

$$Z = \frac{\overline{X} - \mu}{\sigma/\sqrt{n}}.$$

We can rearrange this to equation to read

$$P(\overline{X} + z_\alpha \frac{\sigma}{\sqrt{n}} < \mu < \overline{X} - z_\alpha \frac{\sigma}{\sqrt{n}}) = 1 - 2\alpha \tag{6.12}$$

where we now have μ_{low} and μ_{hi} explicitly.

Example 6.14.: confidence interval on mean, variance known
Samples of dioxin contamination in 36 front yards in St. Louis show a concentration of 6 ppm. Find the 95% confidence interval for the population mean. Assume that the standard deviation is 1.0 ppm.

To solve this, first calculate $\alpha, z_\alpha, z_{1-\alpha}$.

$$1 - 2\alpha = 0.95$$
$$\alpha = 0.025$$
$$z_\alpha = z_{0.025} = -1.96$$
$$z_{1-\alpha} = -z_\alpha = 1.96$$

The z value came from a standard normal table. Alternatively, we can compute this value from MATLAB,

```
>> z = icdf('normal',0.025,0,1)

z =  -1.959963984540055
```

Here we used the inverse cumulative distribution function (icdf) command. Since we have the standard normal PDF, the mean is 0 and the variance is 1. The value of 0.025 corresponds to alpha, the probability.

To get the value of the other limit, we either rely on symmetry, or compute it directly,

```
>> z = icdf('normal',0.975,0,1)

z =  1.959963984540054
```

Note that these values of z are independent of all aspects of the problem except the value of the confidence interval.

Therefore, by equation (6.12)

$$P(6+(-1.96)\frac{1}{\sqrt{36}} < \mu < \overline{X} - (-1.96)\frac{1}{\sqrt{36}} = 1 - 0.05 = 0.95$$

so the 95% confidence interval for the mean is $5.673 < \mu < 6.327$.

Statistic 2. mean, σ unknown: confidence interval

Now usually, we don't know the variance. We have to use our estimate of the variance, s, for σ. In that case, estimating the mean requires the T-distribution. (See previous section.) Let me stress that we do everything exactly as we did before but we use s for σ and use the *t*-distribution instead of the normal distribution. Remember the t-distribution is also symmetric about the origin, so $t_{1-\alpha} = -t_\alpha$. (this means you only have to compute the *t* probability once. Remember, ν=n-1.

$$P(t_\alpha < T < t_{1-\alpha}) = P(t_\alpha < T < -t_\alpha) = 1 - 2\alpha$$

where

$$T = \frac{\overline{X} - \mu}{s/\sqrt{n}}.$$

Just as before, we can rearrange this to equation to read

$$P(\overline{X} + t_\alpha \frac{s}{\sqrt{n}} < \mu < \overline{X} - t_\alpha \frac{s}{\sqrt{n}}) = 1 - 2\alpha \tag{6.13}$$

where we now have μ_{low} and μ_{hi} explicitly.

Example 6.15.: confidence interval on mean, variance unknown

Samples of dioxin contamination in 36 front yards in St. Louis show a concentration of 6 ppm. Find the 95% confidence interval for the population mean. The sample standard deviation, s, was measured to be 1.0.

To solve this, first calculate $\alpha, t_\alpha, t_{1-\alpha}$ for ν = 35.

Sampling and Estimation - 149

$1 - 2\alpha = 0.95$

$\alpha = 0.025$

$t_\alpha = t_{0.025} = -2.03$

$t_{1-\alpha} = -t_\alpha = +2.03$

The t value came from a table of t-distribution values. Alternatively, we can compute this value using MATLAB,

```
>> t = icdf('t',0.025,35)

t =   -2.030107928250342
```

and for the upper limit

```
>> t = icdf('t',0.975,35)

t =    2.030107928250342,
```

which can also be obtained by symmetry. Note that these values of t are independent of all aspects of the problem except the value of the confidence interval and the number of sample points, n.

Therefore, by equation (6.13)

$$P\left(6 - (2.03)\frac{1}{\sqrt{36}} < \mu < \overline{X} + (2.03)\frac{1}{\sqrt{36}}\right) = 1 - 0.05 = 0.95$$

so the 95% confidence interval for the mean is $5.662 < \mu < 6.338$.

You should note that we are a little less confident about the mean when we use the sample variance as the estimate for the population variance, for which the 95% confidence interval for the mean was $5.673 < \mu < 6.327$.

Statistic 3. difference of means, σ known: confidence interval

The exact same derivation that we used above for a single mean can be used for the difference of means. When we the variances of the two samples are known, we have:

$$P\left[(\overline{X}_1 - \overline{X}_2) + z_\alpha \sqrt{\frac{\sigma_1^2}{n_1} + \frac{\sigma_2^2}{n_2}} < (\mu_1 - \mu_2) < (\overline{X}_1 - \overline{X}_2) - z_\alpha \sqrt{\frac{\sigma_1^2}{n_1} + \frac{\sigma_2^2}{n_2}}\right] = 1 - 2\alpha \qquad (6.14)$$

where z is a random variable obeying the standard normal PDF.

Example 6.16.: confidence interval on the difference of means, variances known

Samples of dioxin contamination in 36 front yards in Times Beach, a suburb of St. Louis, show a concentration of 6 ppm with a population variance of 1.0 ppm². Samples of dioxin contamination in 16 front yards in Quail Run, another suburb of St. Louis, show a concentration of 8 ppm with a population variance of 3.0 ppm². Find the 95% confidence interval for the difference of population means. .

To solve this, first calculate $\alpha, z_\alpha, z_{1-\alpha}$.

$1 - 2\alpha = 0.95$

$\alpha = 0.025$

$z_\alpha = z_{0.025} = -1.96$

$z_{1-\alpha} = -z_\alpha = 1.96$

The z value came from a table of standard normal PDF values. Alternatively, we can compute this value from MATLAB,

```
>> z = icdf('normal',0.025,0,1)

z =   -1.959963984540055
```

Therefore, by equation (6.16)

$$P\left[(6-8)-1.96\sqrt{\frac{1}{36}+\frac{3}{16}} < (\mu_1-\mu_2) < (6-8)+1.96\sqrt{\frac{1}{36}+\frac{3}{16}}\right] = 1-2(0.025)$$

$$P[-2.909 < (\mu_1-\mu_2) < -1.091] = 0.95$$

So the 95% confidence interval for the mean is $-2.909 < (\mu_1-\mu_2) < -1.091$.

If we are determining which site is more contaminated, then we are 95% sure that site 2 (Quail Run) is more contaminated by 1 to 3 ppm than site 1, (Times Beach).

Statistic 4. difference of means, σ unknown: confidence interval

When we the variances of the two samples are unknown, we have:

$$P\left[(\overline{X}_1-\overline{X}_2)+t_\alpha\sqrt{\frac{s_1^2}{n_1}+\frac{s_2^2}{n_2}} < (\mu_1-\mu_2) < (\overline{X}_1-\overline{X}_2)+t_{1-\alpha}\sqrt{\frac{s_1^2}{n_1}+\frac{s_2^2}{n_2}}\right] = 1-2\alpha \qquad (6.15)$$

where the number of degrees of freedom for the t-distribution is

Sampling and Estimation - 151

$$v = n_1 + n_2 - 2 \text{ if } \sigma_1 = \sigma_2$$

$$v = \frac{\left(\frac{s_1^2}{n_1} + \frac{s_2^2}{n_2}\right)^2}{\left[\left(\frac{s_1^2}{n_1}\right)^2 / (n_1 - 1)\right] + \left[\left(\frac{s_2^2}{n_2}\right)^2 / (n_2 - 1)\right]} \text{ if } \sigma_1 \neq \sigma_2$$

Example 6.16.: confidence interval on the difference of means, variances unknown

Samples of dioxin contamination in 36 front yards in Times Beach, a suburb of St. Louis, show a concentration of 6 ppm with a sample variance of 1.0 ppm². Samples of dioxin contamination in 16 front yards in Quail Run, another suburb of St. Louis, show a concentration of 8 ppm with a sample variance of 3.0 ppm². Find the 95% confidence interval for the difference of population means. .

To solve this, first calculate $\alpha, t_\alpha, t_{1-\alpha}$.

$$v = \frac{\left(\frac{s_1^2}{n_1} + \frac{s_2^2}{n_2}\right)^2}{\left[\left(\frac{s_1^2}{n_1}\right)^2 / (n_1 - 1)\right] + \left[\left(\frac{s_2^2}{n_2}\right)^2 / (n_2 - 1)\right]} = \frac{\left(\frac{1}{36} + \frac{3}{16}\right)^2}{\left[\left(\frac{1}{36}\right)^2 / (36 - 1)\right] + \left[\left(\frac{3}{16}\right)^2 / (16 - 1)\right]} = 19.59 \approx 20$$

$1 - 2\alpha = 0.95$

$\alpha = 0.025$

$t_\alpha = t_{0.025} = 2.086$

$t_{1-\alpha} = -t_\alpha = -2.086$

The t value came from a table of t-PDF values. Alternatively, we can compute this value using MATLAB,

```
>> t = icdf('t',0.025,20)

t =  -2.085963447265864
```

Therefore, substituting into equation (6.15) yields

$$P\left[(6-8) - 2.086\sqrt{\frac{1}{36} + \frac{3}{16}} < (\mu_1 - \mu_2) < (6-8) + 2.086\sqrt{\frac{1}{36} + \frac{3}{16}}\right] = 1 - 2(0.025)$$

151

$$P[-2.97 < (\mu_1 - \mu_2) < -1.03] = 0.95$$

So the 95% confidence interval for the mean is $-2.97 < (\mu_1 - \mu_2) < -1.03$.

If we are determining which site is more contaminated, then we are 95% sure that site 2 (Quail Run) is more contaminated by 1 to 3 ppm than site 1, (Times Beach).

Statistic 5. variance: confidence interval

The confidence interval of the variance can be estimated in a precisely analogous way, knowing that the statistic

$$\chi^2 = \frac{(n-1)S^2}{\sigma^2} = \sum_{i=1}^{n} \frac{(X_i - \overline{X})^2}{\sigma^2}$$

has a chi-squared distribution with v=n-1 degrees of freedom, $f_{\chi^2}(\chi^2; n-1)$. So

$$P\left[\frac{(n-1)s^2}{\chi^2_{1-\alpha}} < \sigma^2 < \frac{(n-1)s^2}{\chi^2_{\alpha}}\right] = 1 - 2\alpha \tag{6.16}$$

Perversely, the tables of the critical values for the χ^2 distribution, have defined α to be 1-α, so the indices have to be switched when using the table.

$$P\left[\frac{(n-1)s^2}{\chi^2_{\alpha}} < \sigma^2 < \frac{(n-1)s^2}{\chi^2_{1-\alpha}}\right] = 1 - 2\alpha \quad \text{when using the } \chi^2 \text{ critical values table only!}$$

If you get confused, just remember that the upper limit must be greater than the lower limit. Remember also that the $f_{\chi^2}(\chi^2; n-1)$ is not symmetric about the origin, so we cannot use the symmetry arguments used for the confidence intervals for functions of the mean.

Example 6.17.: variance

Samples of dioxin contamination in 16 front yards in St. Louis show a concentration of 6 ppm. Find the 95% confidence interval for the population mean. The sample standard deviation, *s*, was measured to be 1.0.

To solve this, first calculate $\alpha, \chi^2_{\alpha}, \chi^2_{1-\alpha}$.

For v = n – 1 = 15, we have

$1 - 2\alpha = 0.95$

$\alpha = 0.025$

$\chi_\alpha^2 = \chi_{0.025}^2 = 27.488$

$\chi_{1-\alpha}^2 = \chi_{0.975}^2 = 6.262$

The t value came from a table of χ^2-distribution values. Alternatively, we can compute this value using MATLAB,

```
>> chi2 = icdf('chi2',0.025,15)

chi2 =   6.262137795043251
```

and

```
>> chi2 = icdf('chi2',0.975,15)

chi2 =  27.488392863442972
```

Therefore, substituting into equation (6.16) yields

$$P\left[\frac{(16-1)1.0}{27.488} < \sigma^2 < \frac{(16-1)1.0}{6.262}\right] = 1 - 2(0.025)$$

$$P(0.5457 < \sigma^2 < 2.395) = 0.95$$

So the 95% confidence interval for the mean is $0.5457 < \sigma^2 < 2.395$.

Statistic 6. ratio of variances: confidence interval (p. 253)

The ratio of two population variances can be estimated in a precisely analogous way, knowing that the statistic

$$F = \frac{S_1^2/\sigma_1^2}{S_2^2/\sigma_2^2} = \frac{S_1^2 \sigma_2^2}{S_2^2 \sigma_1^2}$$

follows the F-distribution with $v_1 = n_1 - 1$ and $v_2 = n_2 - 1$ degrees of freedom. Remember, the F-distribution has a symmetry, $f_{1-\alpha/2}(v_1, v_2) = \dfrac{1}{f_{\alpha/2}(v_2, v_1)}$. This symmetry relation is essential if one is to use tables for the critical value of the F-distribution. It is not essential if one uses MATLAB commands.

If one is computing the cumulative PDF for the f distribution, then one simply, rearranges this equation for $\frac{\sigma_1^2}{\sigma_2^2}$

$$\frac{\sigma_2^2}{\sigma_1^2} = F \frac{S_2^2}{S_1^2}$$

$$\frac{\sigma_1^2}{\sigma_2^2} = \frac{1}{F} \frac{S_1^2}{S_2^2}$$

$$P\left[\frac{S_1^2}{S_2^2} \frac{1}{f_{1-\alpha}(v_1, v_2)} < \frac{\sigma_1^2}{\sigma_2^2} < \frac{S_1^2}{S_2^2} \frac{1}{f_\alpha(v_1, v_2)} \right] = 1 - 2\alpha \qquad (6.17)$$

One notes that the order of the limits has changed here, since as $\frac{\sigma_1^2}{\sigma_2^2}$ goes up, F goes down. In any case, the lower limit must be smaller than the upper limit. If one chooses to use tables of critical values, one must take into account two idiosyncrasies of the procedure. First, as was the case with the t and chi-squared distributions, the table provide the probability that f is greater than a value, not the cumulative PDF, which is the probability that f is less than a value. Second, the tables only provide data for small values of α. Therefore, we must eliminate all instances of 1-α., using a symmetry relation. The result is

$$P\left[\frac{S_1^2}{S_2^2} \frac{1}{f_\alpha(v_1, v_2)} < \frac{\sigma_1^2}{\sigma_2^2} < \frac{S_1^2}{S_2^2} f_\alpha(v_2, v_1) \right] = 1 - 2\alpha \quad \text{when using the tables only!}$$

Example 6.18.: confidence interval on the ratio of variances
Samples of dioxin contamination in 20 front yards in Times Beach, a suburb of St. Louis, show a concentration of 6 ppm with a sample variance of 1.0 ppm². Samples of dioxin contamination in 16 front yards in Quail Run, another suburb of St. Louis, show a concentration of 8 ppm with a sample variance of 3.0 ppm². Find the 90% confidence interval for the difference of population means. .
To solve this, first calculate $\alpha, F_\alpha, F_{1-\alpha}$, with $v_1 = n_1 - 1 = 19$ and $v_2 = n_2 - 1 = 15$

$1 - 2\alpha = 0.90$

$\alpha = 0.05$

We can compute the f probabilities using MATLAB,

Sampling and Estimation - 155

```
>> f = icdf('f',0.05,19,15)

f =    0.447614966503185
```

and

```
>> f = icdf('f',0.95,19,15)

f =    2.339819281665456
```

Substituting into equation (6.16) yields

$$P\left[\frac{1}{3}\frac{1}{2.3398} < \frac{\sigma_1^2}{\sigma_2^2} < \frac{1}{3}\frac{1}{0.4476}\right] = 1 - 2(0.05)$$

$$P\left[0.1425 < \frac{\sigma_1^2}{\sigma_2^2} < 0.7447\right] = 0.90$$

Alternatively, we can use the table of critical values

$$F_\alpha = F_{0.05} = F_{0.05}(v_1 = 19, v_2 = 15) \approx F_{0.05}(v_1 = 20, v_2 = 15) = 2.33$$
$$F_{0.05}(v_1 = 15, v_2 = 19) = 2.23$$

$$P\left[\frac{1}{3}\frac{1}{2.33} < \frac{\sigma_1^2}{\sigma_2^2} < \frac{1}{3}2.23\right] = 1 - 2(0.05)$$

$$P\left[0.1431 < \frac{\sigma_1^2}{\sigma_2^2} < 0.7433\right] = 0.90$$

So the 90% confidence interval for the mean is $0.1425 < \frac{\sigma_1^2}{\sigma_2^2} < 0.7447$.

If we are determining which site has a greater variance of contamination levels then we are 90% sure that site 2 (Quail Run) has more variance by a factor of 1.3 to 7.0.

6.5. Problems

We intend to purchase a liquid as a raw material for a material we are designing. Two vendors offer us samples of their product and a statistic sheet. We run the samples in our own labs and come up with the following data:

Vendor 1		Vendor 2	
sample #	outcome	sample #	outcome
1	2.3	1	2.49
2	2.49	2	1.98
3	2.05	3	2.18
4	2.4	4	2.36
5	2.18	5	2.47
6	2.12	6	2.36
7	2.38	7	1.82
8	2.39	8	1.88
9	2.4	9	1.87
10	2.46	10	1.87
11	2.19		
12	2.04		
13	2.43		
14	2.34		
15	2.19		
16	2.12		

Vendor Specification Claims:
Vendor 1: $\mu = 2.0$ and $\sigma^2 = 0.05$, $\sigma = 0.2236$
Vendor 2: $\mu = 2.3$ and $\sigma^2 = 0.12$, $\sigma = 0.3464$

Sample statistics, based on the data provided in the table above.

$$n_1 = 16 \quad \bar{x}_1 = \frac{1}{16}\sum_{i=1}^{16} x_i = 2.280 \quad s_1^2 = \frac{1}{16}\sum_{i=1}^{16}(x_i - \bar{x}_1)^2 = 0.0229 \quad s_1 = 0.1513$$

$$n_2 = 10 \quad \bar{x}_2 = \frac{1}{10}\sum_{i=1}^{10} x_i = 2.128 \quad s_2^2 = \frac{1}{10}\sum_{i=1}^{10}(x_i - \bar{x}_2)^2 = 0.0744 \quad s_2 = 0.2728$$

Problem 6.1.
Determine a 95% confidence interval on the mean of sample 1. Use the value of the population variance given. Is the given population mean legitimate?

Problem 6.2.
Determine a 95% confidence interval on the difference of means between samples 1 and 2. Use the values of the population variance given. Is the difference between the given population means legitimate?

Problem 6.3.
Determine a 95% confidence interval on the mean of sample 1. Assume the given values of the population variances are suspect and not to be trusted. Is the given population mean legitimate?

Problem 6.4.
Determine a 95% confidence interval on the difference of means between samples 1 and 2. Assume the given values of the population variances are suspect and not to be trusted. Is the difference between the given population means legitimate?

Problem 6.5.
Determine a 95% confidence interval on the variance of sample 1. Is the given population variance legitimate?

Problem 6.6.
Determine a 98% confidence interval on the ratio of variance of samples 1 & 2. Is the ratio of the given population variances legitimate?

Appendix I. Review Handout for Discrete Probability Distribution Functions

A.1. Discrete uniform $$f(x;k) = \frac{1}{k}$$	k = number of elements in sample space x = outcome is one distinct element
A.2. Binomial $$b(x;n,p) = \binom{n}{x} p^x q^{n-x} \text{ for } x = 0,1,2\ldots n$$	n = # of (independent, repeated, with replacement, only 2 outcomes) Bernoulli trials p = probability of success on one trial $q = 1 - p$ = probability of failure on one trial x = # of successes
A.3. Multinomial $$m(\{x\};n,\{p\};k) = \binom{n}{x_1, x_2 \ldots x_k} \prod_{i=1}^{k} p_i^{x_i}$$	k = # of different types of outcomes n = # of (independent, repeated, with replacement) Bernoulli trials p_i = probability of type i success on one trial x_i = # of successes of type i
A.4. Hypergeometric $$h(x;N,n,k) = \frac{\binom{k}{x}\binom{N-k}{n-x}}{\binom{N}{n}} \text{ for } x = 0,1,2\ldots n$$	N = # of elements in population n = # of elements in sample, drawn without replacement k = # of outcomes labeled success in population x = # of successes in sample
A.5. Multivariate Hypergeometric $$h_m(\{x\};N,n,\{a\};k) = \frac{\binom{a_1}{x_1}\binom{a_2}{x_2}\binom{a_3}{x_3}\cdots\binom{a_k}{x_k}}{\binom{N}{n}}$$	k = # of different types of outcomes N = # of elements in population n = # of elements in sample, drawn without replacement a_i = # of outcomes labeled success of type i in population x_i = # of successes of type i in sample
A.6. Negative Binomial $$b^*(x;k,p) = \binom{x-1}{k-1} p^k q^{x-k}$$ $$\text{for } x = k, k+1, k+2\ldots$$	x = # of (independent, repeated, with replacement, only 2 outcomes) Bernoulli trials p = probability of success on one trial $q = 1 - p$ = probability of failure on one trial k = # of successes
A.7. Geometric $$g(x;p) = pq^{x-1} \text{ for } x = 1,2,3\ldots$$	x = # of (independent, repeated, with replacement, only 2 outcomes) Bernoulli trials p = probability of success on one trial $q = 1 - p$ = probability of failure on one trial
A.8. Poisson $$p(x;\lambda t) = \frac{e^{-\lambda t}(\lambda t)^x}{x!} \text{ for } x = 0,1,2\ldots$$	t = the size of the interval λ = the rate of the occurrence of the outcome x = the number of outcomes occurring in interval t.

Appendix II. Review Handout for Continuous Probability Density Functions

B.1. Continuous uniform $$f(x; A, B) = \begin{cases} \dfrac{1}{B-A} & \text{for } A \leq x \leq B \\ 0 & \text{otherwise} \end{cases}$$	A = lower limit of random variable B = upper limit of random variable x = outcome of uniform selection
B.2. Normal $$f(x; \mu, \sigma) = \dfrac{1}{\sqrt{2\pi}\sigma} e^{-\frac{1}{2}\left(\frac{x-\mu}{\sigma}\right)^2}$$ standard normal PDF uses $z = \dfrac{x-\mu}{\sigma}$	μ = population mean σ = population standard deviation x = random variable of normal PDF
B.3. Gamma $$f_\Gamma(x; \alpha, \beta) = \begin{cases} \dfrac{1}{\beta^\alpha \Gamma(\alpha)} x^{\alpha-1} e^{-x/\beta} & \text{for } x > 0 \\ 0 & \text{elsewhere} \end{cases}$$	α = # of events β = mean time to failure, or mean time between events x = time of interest $\Gamma(\alpha)$ = Gamma Function
B.4. Exponential (gamma with $\alpha = 1$) $$f_e(x; \beta) = \begin{cases} \dfrac{1}{\beta} e^{-x/\beta} & \text{for } x > 0 \\ 0 & \text{elsewhere} \end{cases}$$	β = mean time to failure, or mean time between events x = time of interest
B.5. Chi-squared $$f_{\chi^2}(x; v) = \begin{cases} \dfrac{1}{2^{v/2} \Gamma(v/2)} x^{v/2-1} e^{-x/2} & \text{for } x > 0 \\ 0 & \text{elsewhere} \end{cases}$$	$v = n-1$ = degrees of freedom, x = random variable $\alpha = 1 - F_{\chi^2}(x; v) = P(X > x)$
B.6. t-distribution $$f(t) = \dfrac{\Gamma[(v+1)/2]}{\Gamma(v/2)\sqrt{\pi v}} \left(1 + \dfrac{t^2}{v}\right)^{-\frac{v+1}{2}} \text{ for } -\infty < t < \infty$$	$v = n-1$ = degrees of freedom, x = random variable $\alpha = 1 - F_t(x; v) = P(X > x)$
B.7. F-distribution $$h_f(f; v_1, v_2) = \begin{cases} \dfrac{\Gamma\left(\dfrac{v_1+v_2}{2}\right)\left(\dfrac{v_1}{v_2}\right)^{\frac{v_1}{2}} f^{\frac{v_1}{2}-1}}{\Gamma\left(\dfrac{v_1}{2}\right)\Gamma\left(\dfrac{v_2}{2}\right)\left(1 + \dfrac{v_1}{v_2} f\right)^{\frac{v_1+v_2}{2}}} & \text{for } f > 0 \\ 0 & \text{elsewhere} \end{cases}$$	$v_1 = n_1 - 1$ = degrees of freedom of variable 1 $v_2 = n_2 - 1$ = degrees of freedom of variable 2 $\alpha = 1 - F_F(x; v) = P(X > x)$

Appendix III. Review Handout for Statistics

μ = population mean, σ^2 = population variance, $\bar{x}$ = sample mean, s^2 = sample variance
Confidence Interval = $1 - 2\alpha$, in all cases, α represents area to left (less than probability).

A. Mean, population variance known
Use standard normal distribution.

transformation: $Z = \dfrac{\bar{X} - \mu}{\sigma / \sqrt{n}}$

symmetry: $z_{1-\alpha} = -z_\alpha$

confidence interval: $P(\bar{X} + z_\alpha \dfrac{\sigma}{\sqrt{n}} < \mu < \bar{X} + z_{1-\alpha} \dfrac{\sigma}{\sqrt{n}}) = 1 - 2\alpha$

B. Mean, population variance unknown
Use t-distribution.

transformation: $T = \dfrac{\bar{X} - \mu}{s / \sqrt{n}}$

symmetry: $t_{1-\alpha} = -t_\alpha$

parameters: $v = n - 1$

confidence interval: $P(\bar{X} + t_\alpha \dfrac{s}{\sqrt{n}} < \mu < \bar{X} + t_{1-\alpha} \dfrac{s}{\sqrt{n}}) = 1 - 2\alpha$

C. Difference of Means, population variance known
Use standard normal distribution.

transformation: $Z = \dfrac{(\bar{X}_1 - \bar{X}_2) - (\mu_1 - \mu_2)}{\sqrt{\left(\dfrac{\sigma_1^2}{n_1}\right) + \left(\dfrac{\sigma_2^2}{n_2}\right)}}$

symmetry: $z_{1-\alpha} = -z_\alpha$

confidence interval:

$$P\left[(\bar{X}_1 - \bar{X}_2) + z_\alpha \sqrt{\dfrac{\sigma_1^2}{n_1} + \dfrac{\sigma_2^2}{n_2}} < (\mu_1 - \mu_2) < (\bar{X}_1 - \bar{X}_2) + z_{1-\alpha} \sqrt{\dfrac{\sigma_1^2}{n_1} + \dfrac{\sigma_2^2}{n_2}}\right] = 1 - 2\alpha$$

D. Difference of Means, population variance unknown
Use t-distribution.

transformation: $T = \dfrac{(\overline{X}_1 - \overline{X}_2) - (\mu_1 - \mu_2)}{\sqrt{\left(\dfrac{s_1^2}{n_1}\right) + \left(\dfrac{s_2^2}{n_2}\right)}}$

symmetry: $t_{1-\alpha} = -t_\alpha$

parameters: $v = n_1 + n_2 - 2$ if $\sigma_1 = \sigma_2$

parameters: $v = \dfrac{\left(\dfrac{s_1^2}{n_1} + \dfrac{s_2^2}{n_2}\right)^2}{\left[\left(\dfrac{s_1^2}{n_1}\right)^2 / (n_1 - 1)\right] + \left[\left(\dfrac{s_2^2}{n_2}\right)^2 / (n_2 - 1)\right]}$ if $\sigma_1 \neq \sigma_2$

confidence interval:

$$P\left[(\overline{X}_1 - \overline{X}_2) + t_\alpha \sqrt{\dfrac{s_1^2}{n_1} + \dfrac{s_2^2}{n_2}} < (\mu_1 - \mu_2) < (\overline{X}_1 - \overline{X}_2) + t_{1-\alpha} \sqrt{\dfrac{s_1^2}{n_1} + \dfrac{s_2^2}{n_2}}\right] = 1 - 2\alpha$$

E. variance, population variance unknown
Use chi-squared distribution.

transformation: $\chi^2 = \dfrac{(n-1)s^2}{\sigma^2}$

no symmetry

parameters: $v = n - 1$

confidence interval: $P\left[\dfrac{(n-1)s^2}{\chi^2_{1-\alpha}} < \sigma^2 < \dfrac{(n-1)s^2}{\chi^2_\alpha}\right] = 1 - 2\alpha$

F. ratio of variances, population variances unknown
Use F-distribution.

transformation: $F = \dfrac{S_1^2 / \sigma_1^2}{S_2^2 / \sigma_2^2} = \dfrac{S_1^2 \sigma_2^2}{S_2^2 \sigma_1^2}$

parameters: $v_1 = n_1 - 1$, $v_2 = n_2 - 1$

symmetry: $\dfrac{1}{f_{1-\alpha}(v_1, v_2)} = f_\alpha(v_2, v_1)$

confidence interval: $P\left[\dfrac{S_1^2}{S_2^2} \dfrac{1}{f_{1-\alpha}(v_1, v_2)} < \dfrac{\sigma_1^2}{\sigma_2^2} < \dfrac{S_1^2}{S_2^2} \dfrac{1}{f_\alpha(v_1, v_2)}\right] = 1 - 2\alpha$

Appendix IV. Summary of MATLAB Statistics Commands & References

I. Discrete Distributions
The official online document on discrete distributions is located at the following URL:
http://www.mathworks.com/help/stats/discrete-distributions.html

MATLAB contains intrinsic functions describing all discrete PDFs studied in this course except the multivariate hypergeometric.
- Uniform Distribution (Discrete)
- Binomial Distribution
- Multinomial Distribution
- Hypergeometric Distribution
- Negative Binomial Distribution
- Geometric Distribution
- Poisson Distribution

II. Continuous Distributions
The official online document on continuous distributions is located at the following URL:
http://www.mathworks.com/help/stats/continuous-distributions.html

MATLAB contains intrinsic functions describing all continuous PDFs studied in this course and many more.
- Uniform Distribution (Continuous)
- Normal Distribution
- Gamma Distribution
- Exponential Distribution
- Chi-Squared Distribution
- t Distribution
- F Distribution

III. Calculating the value of the PDF
The official online document on the pdf function is located at the following URL:
http://www.mathworks.com/help/stats/pdf.html

MATLAB contains intrinsic functions for calculating values of PDFs. For discrete PDFs, these values are probabilities. For continuous PDFs, these are values of the integrand. To calculate the value of PDF, one can use the pdf function.

```
Y = pdf(name,X,A)
Y = pdf(name,X,A,B)
Y = pdf(name,X,A,B,C)
```

where x is the random variable and A, B and C are parameters.
The values of the name of the pdf and the parameters are provided in the following table.

PDF Table for MATLAB

name	Distribution	Input Parameter A	Input Parameter B	Input Parameter C
'unid' or 'Discrete Uniform'	Uniform Distribution (Discrete)	N: maximum observable value	—	—
'bino' or 'Binomial'	Binomial Distribution	n: number of trials	p: probability of success for each trial	—
'hyge' or 'Hypergeometric'	Hypergeometric Distribution	M: size of the population	K: number of items with the desired characteristic in the population	n: number of samples drawn
'nbin' or 'Negative Binomial'	Negative Binomial Distribution	r: number of successes	p: probability of success in a single trial	—
'geo' or 'Geometric'	Geometric Distribution	p: probability parameter	—	—
'poiss' or 'Poisson'	Poisson Distribution	λ: mean	—	—
'unif' or 'Uniform'	Uniform Distribution (Continuous)	a: lower endpoint (minimum)	b: upper endpoint (maximum)	—
'norm' or 'Normal'	Normal Distribution	μ: mean	σ: standard deviation	—
'gam' or 'Gamma'	Gamma Distribution	a: shape parameter	b: scale parameter	—
'exp' or 'Exponential'	Exponential Distribution	μ: mean	—	—
'chi2' or 'Chisquare'	Chi-Square Distribution	v: degrees of freedom	—	—
't' or 'T'	Student's t Distribution	v: degrees of freedom	—	—
'f' or 'F'	F Distribution	$v1$: numerator degrees of freedom	$v2$: denominator degrees of freedom	—

Two examples of the usage of the pdf function are given below.

```
>> f = pdf('Normal',-2,0,1)
f = 0.053990966513188
>> p = pdf('Poisson',4,5)
p = 0.175467369767851
```

IV. Calculating the value of the Cumulative PDF

The official online document on the cdf function is located at the following URL:
http://www.mathworks.com/help/stats/cdf.html

MATLAB contains intrinsic functions for calculating values of the cumulative probability of PDFs. To calculate the value of the cumulative probability of the PDF, one can use the cdf function.

```
y = cdf('name',x,A)
y = cdf('name',x,A,B)
y = cdf('name',x,A,B,C)
```

Several examples of the usage of the pdf function are given below.

(a) Normal distribution - $P(z < 1.43; \mu = 5, \sigma = 3)$

```
>> p = cdf('Normal',1.43,5,3)

p = 0.117023196023109
```

(b) Normal distribution - $P(z > 1.43; \mu = 5, \sigma = 3)$

```
>> p = 1 - cdf('Normal',1.43,5,3)

p =   0.882976803976891
```

(c) Normal distribution - $P(4.5 < z < 6.5; \mu = 5, \sigma = 3)$

```
>> p = cdf('Normal',6.5,5,3) - cdf('Normal',4.5,5,3)

p = 0.257646293884917
```

(d) Binomial distribution - $P(x \leq 2; n = 5, p = 0.3)$

```
>> p = cdf('Binomial',2,5,0.3)

p = 0.836920000000000
```

(e) Binomial distribution - $P(x \geq 2; n = 5, p = 0.3)$

```
>> p = 1 - cdf('Binomial',1,5,0.3)

p = 0.471780000000000
```

(f) Binomial distribution - $P(1 \leq x \leq 2; n = 5, p = 0.3)$

```
>> p = cdf('Binomial',2,5,0.3) - cdf('Binomial',0,5,0.3)

p = 0.668850000000000
```

V. Inverse Problem – given p, find z (or x)

The official online document on the icdf function is located at the following URL:
http://www.mathworks.com/help/stats/icdf.html

MATLAB contains intrinsic functions for calculating values random variables corresponding to given cumulative probabilities for various PDFs. This is the "given z, find p" problem.

```
Y = icdf(name,X,A)
Y = icdf(name,X,A,B)
Y = icdf(name,X,A,B,C)
```

(a) Normal distribution - $P(z < Z; \mu = 5, \sigma = 3) = 0.05$, find Z

```
>> z = icdf('Normal',0.05,5,3)

z = 0.065439119145582
```

(b) Normal distribution - $P(z > Z; \mu = 5, \sigma = 3) = 0.05$

Use the probability $1 - 0.05 = 0.95$ for this "greater than" probability.

```
>> z = icdf('Normal',0.95,5,3)

z = 9.934560880854416
```

VI. Other notes

Some distributions are more common than others and have specific functions named after them. For example, for the normal distribution, there are two ways to calculate the cumulative PDF,

```
>> p = cdf('Normal',1.43,5,3)
>> p = normcdf(1.43,5,3)
```

However, it appears that these specific functions are simply redundant function names. They don't exist for all PDFs or all cumulative PDFs and there are no special functions for the inverse cumulative PDF function (icdf).

Appendix V. Derivation of the fact that the distribution of the mean is normal

Consider taking n samples from a population characterized by mean, μ, and variance, σ^2. The sample mean is given by $\bar{x}$. We define a moment generating function for a continuous PDF to be:

$$M_x(t) = \mu_{e^{tx}} = E\left[e^{tx}\right] = \int_{-\infty}^{\infty} e^{tx} f(x)dx$$

From this definition, and using the rules of linear operation, we can show that

$$M_{x+a}(t) = \mu_{e^{t(x+a)}} = E\left[e^{t(x+a)}\right] = \int_{-\infty}^{\infty} e^{t(x+a)} f(x)dx = \int_{-\infty}^{\infty} e^{tx} e^{ta} f(x)dx = e^{ta} \int_{-\infty}^{\infty} e^{tx} f(x)dx$$

$$M_{x+a}(t) = e^{ta} M_x(t) \tag{V.1}$$

$$M_{ax}(t) = \mu_{e^{tax}} = E\left[e^{tax}\right] = \int_{-\infty}^{\infty} e^{tax} f(x)dx = M_x(at)$$

$$M_{ax}(t) = M_x(at) \tag{V.2}$$

$$M_{x_1+x_2+x_3+\ldots+x_n}(t) = \mu_{e^{t(x_1+x_2+x_3+\ldots+x_n)}} = E\left[e^{t(x_1+x_2+x_3+\ldots+x_n)}\right] = \int_{-\infty}^{\infty} e^{t(x_1+x_2+x_3+\ldots+x_n)} f(x)dx$$

$$M_{x_1+x_2+x_3+\ldots+x_n}(t) = M_{x_1}(t) M_{x_2}(t) M_{x_3}(t) \ldots M_{x_n}(t) \tag{V.3}$$

So that

$$M_{(\bar{x}-\mu)/(\sigma/\sqrt{n})}(t) = \mu_{e^{t[(\bar{x}-\mu)/(\sigma/\sqrt{n})]}} = E\left[e^{t[(\bar{x}-\mu)/(\sigma/\sqrt{n})]}\right] = \int_{-\infty}^{\infty} e^{t[(\bar{x}-\mu)/(\sigma/\sqrt{n})]} f(\bar{x})d\bar{x}$$

$$M_{(\bar{x}-\mu)/(\sigma/\sqrt{n})}(t) = \int_{-\infty}^{\infty} e^{t\bar{x}/(\sigma/\sqrt{n})} e^{-t\mu/(\sigma/\sqrt{n})} f(\bar{x})d\bar{x} = e^{-t\mu\sqrt{n}/\sigma} \int_{-\infty}^{\infty} e^{t\bar{x}\sqrt{n}/\sigma} f(\bar{x})d\bar{x}$$

$$M_{(\bar{x}-\mu)/(\sigma/\sqrt{n})}(t) = e^{-t\mu\sqrt{n}/\sigma} M_{\bar{x}}\left(\frac{t\sqrt{n}}{\sigma}\right)$$

Appendix V - 167

Now consider that $\bar{X} = \frac{1}{n}\sum_{i=1}^{n} X_i$, so

$$M_{\bar{x}}\left(\frac{t\sqrt{n}}{\sigma}\right) = M_{\frac{1}{n}\sum_{i=1}^{n}X_i}\left(\frac{t\sqrt{n}}{\sigma}\right) = \int_{-\infty}^{\infty} e^{t\sqrt{n}/\sigma \cdot \frac{1}{n}\sum_{i=1}^{n}X_i} f(x)dx = \int_{-\infty}^{\infty} e^{t/(\sigma\sqrt{n})\sum_{i=1}^{n}X_i} f(x)dx$$

$$M_{\bar{x}}(t\sqrt{n}/\sigma) = \int_{-\infty}^{\infty} e^{t/(\sigma\sqrt{n})\sum_{i=1}^{n}X_i} f(x)dx = M_{x_1}\left(\frac{t}{\sigma\sqrt{n}}\right)M_{x_2}\left(\frac{t}{\sigma\sqrt{n}}\right)...M_{x_n}\left(\frac{t}{\sigma\sqrt{n}}\right)$$

Since there isn't anything intrinsic that distinguishes one X_i from another, we can write

$$M_{\bar{x}}(t\sqrt{n}/\sigma) = \left[M_x\left(\frac{t}{\sigma\sqrt{n}}\right)\right]^n$$

If we substitute this back into our original equation, we have

$$M_{\frac{(\bar{x}-\mu)}{(\sigma/\sqrt{n})}}(t) = e^{-t\mu\sqrt{n}/\sigma} M_{\bar{x}}\left(\frac{t\sqrt{n}}{\sigma}\right) = e^{-t\mu\sqrt{n}/\sigma}\left[M_x\left(\frac{t}{\sigma\sqrt{n}}\right)\right]^n$$

We take the natural log of both sides:

$$\ln\left[M_{\frac{(\bar{x}-\mu)}{(\sigma/\sqrt{n})}}(t)\right] = \frac{-t\mu\sqrt{n}}{\sigma} + n\ln\left[M_x\left(\frac{t}{\sigma\sqrt{n}}\right)\right]$$

We expand $M_x\left(\frac{t}{\sigma\sqrt{n}}\right)$ as an infinite series in powers of t about $t=0$.

$$M_x\left(\frac{t}{\sigma\sqrt{n}}\right) = 1 + v_1 t + v_2 \frac{t^2}{2!} + v_3 \frac{t^3}{3!} + ... + v_r \frac{t^r}{r!} + ...$$

where

$$v_i = \left.\frac{d^i M_x\left(\frac{t}{\sigma\sqrt{n}}\right)}{dt^i}\right|_{t=0}$$

We can write this as

$$M_x\left(\frac{t}{\sigma\sqrt{n}}\right) = 1 + v(t)$$

where $v(t)$ is an infinite series in t. For very large sample sizes, n

$$\lim_{n\to\infty} \ln\left[M_{\frac{(\bar{x}-\mu)}{(\sigma/\sqrt{n})}}(t)\right] = \lim_{n\to\infty} \ln[1 + v(t)] = \frac{t^2}{2}.$$

This can be shown by expanding the natural log in a Maclaurin series. For the present purposes, we will take this step given above on faith. Then, we have

$$\lim_{n\to\infty} M_{\frac{(\bar{x}-\mu)}{(\sigma/\sqrt{n})}}(t) = e^{\frac{t^2}{2}}$$

So the first moment of $Z = \dfrac{\bar{X}-\mu}{\sigma/\sqrt{n}}$ in the limit of large n is $e^{\frac{t^2}{2}}$

Well, let's find what the moment of the random variable, z, would be if it follows the normal distribution. The PDF of the normal distribution is

$$f(x;\mu,\sigma) = \frac{1}{\sqrt{2\pi}\sigma} e^{-\frac{1}{2}\left(\frac{x-\mu}{\sigma}\right)^2}$$

$$M_x(t) = \mu_{e^{tx}} = E\left[e^{tx}\right] = \int_{-\infty}^{\infty} e^{tx} f(x)dx = \int_{-\infty}^{\infty} e^{tx} \frac{1}{\sqrt{2\pi}\sigma} e^{-\frac{1}{2}\left(\frac{x-\mu}{\sigma}\right)^2} dx$$

$$M_x(t) = \int_{-\infty}^{\infty} e^{tx} \frac{1}{\sqrt{2\pi}\sigma} e^{\frac{-x^2+2x\mu-\mu^2}{2\sigma^2}} dx = \int_{-\infty}^{\infty} e^{\frac{2tx\sigma^2}{2\sigma^2}} \frac{1}{\sqrt{2\pi}\sigma} e^{\frac{-x^2+2x\mu-\mu^2}{2\sigma^2}} dx$$

Appendix V - 169

$$M_x(t) = \int_{-\infty}^{\infty} \frac{1}{\sqrt{2\pi}\sigma} e^{\frac{2tx\sigma^2 - x^2 + 2x\mu - \mu^2}{2\sigma^2}} dx = \int_{-\infty}^{\infty} \frac{1}{\sqrt{2\pi}\sigma} e^{\frac{-x^2 + 2(t\sigma^2 + \mu)x - \mu^2}{2\sigma^2}} dx$$

We complete the square in the exponent:

$$-x^2 + 2(t\sigma^2 + \mu)x - \mu^2 = [x - (t\sigma^2 + \mu)]^2 - 2\mu t\sigma^2 - t^2\sigma^4$$

$$M_x(t) = \int_{-\infty}^{\infty} \frac{1}{\sqrt{2\pi}\sigma} e^{-\frac{[x-(t\sigma^2+\mu)]^2 - 2\mu t\sigma^2 - t^2\sigma^4}{2\sigma^2}} dx = e^{\mu t + \frac{t^2\sigma^2}{2}} \int_{-\infty}^{\infty} \frac{1}{\sqrt{2\pi}\sigma} e^{-\frac{[x-(t\sigma^2+\mu)]^2}{2\sigma^2}} dx$$

Let $w = \frac{[x - (t\sigma^2 + \mu)]}{\sigma}$ so that $dw = \frac{dx}{\sigma}$ and

$$M_x(t) = e^{\mu t + \frac{t^2\sigma^2}{2}} \int_{-\infty}^{\infty} \frac{1}{\sqrt{2\pi}} e^{-\frac{w^2}{2}} dw = e^{\mu t + \frac{t^2\sigma^2}{2}} (1) = e^{\mu t + \frac{t^2\sigma^2}{2}}$$

So that the first moment generating function of the standard normal PDF is

$$M_x(t) = e^{\frac{t^2}{2}}$$

If we compare this moment generating function with that obtained for

$$\lim_{n\to\infty} M_{\frac{(\bar{x}-\mu)}{(\sigma/\sqrt{n})}}(t) = e^{\frac{t^2}{2}}$$

we find that they are the same in the limit of large n. Since there is a one-to-one correspondence between PDFs and moment-generating functions, we see that the PDF for $Z = \frac{\bar{X} - \mu}{\sigma/\sqrt{n}}$ is the standard normal PDF.

References

Montgomery, D.C., Runger, D.C., <u>Applied Statistics and Probability for Engineers</u>, 2nd Edition, John Wiley & Sons, New York, 1999.

Walpole, R.E., Myers, R.H, Myers, S.L., <u>Probability and Statistics for Engineers and Scientists</u>, 6th Edition, Prentice Hall, Upper Saddle River, New Jersey, 1998.

CPSIA information can be obtained
at www.ICGtesting.com
Printed in the USA
LVHW050439310720
661978LV00007B/356